Praise for *Censored Landscapes*

"This is perhaps the most powerful indictment against animal agriculture I've ever encountered. It gave me an eerie feeling of deafening silence, while inviting a sliver of hope."—**Dr. Jonathan Balcombe**, author of *Super Fly, What a Fish Knows, Second Nature,* and *Pleasurable Kingdom*

"*Censored Landscapes* is an eye-opening testament to the profound impact of art in unveiling the hidden truths of our world. Among the most egregious and disturbing unseen truths concerns the abusive mistreatment of so-called food animals who, despite being highly sentient, feeling beings, are treated as if they are unfeeling objects whose only purpose in life is to provide food for carnivorous humans. It's a matter of *who* is for dinner, not *what* is for dinner. Through powerful photographs, insightful text, and detailed research, Isabella La Rocca González exposes an industry built on webs of social inequality and injustices and the violent exploitation of innocent animals. With meticulous attention to detail and poignant storytelling, *Censored Landscapes* invites readers to confront uncomfortable truths with empathy and compassion. In a world grappling with ethical dilemmas and ecological crises, this timely work demonstrates the transformative power of art and advocacy to provoke thought and ignite action. *Censored Landscapes* not only informs but also draws the connection between animal agriculture and every facet of our lives. This timely work is a call to action for a more just and equitable world." —**Marc Bekoff**, PhD, University of Colorado, Boulder, author of *The Animals' Agenda: Freedom, Compassion, and Coexistence in the Human Age,* and *The Emotional Lives of Animals: A Leading Scientist Explores Animal Joy, Sorrow, and Empathy—and Why They Matter*

"The architecture of power and control is so blatant and banal, it's almost invisible. Almost, but not to this photographer. For the animals, the innocent, the gentle, these places of terror and grief scar the land. The stench and cries are carried by the wind and water, as they have memory. Roads, like arteries, end at a slaughterhouse in America, from the slaughterhouse to your house. You consume their terror. The portraits of the few refugees who have found shelter, a home, are poignantly contrasted with the trillion farmed animals born into mechanized death. Meticulously documented facts and essays in this book support urgency to change, for the road we are on is coming to an end."—**Sue Coe**, artist and author of *The Animals' Manifesto, The Ghosts of Our Meat, Dead Meat,* and *Cruel*

"Kept at a distance by belligerent guards, Isabella La Rocca González has produced a suite of photographs of the death factories that, even from afar, give off a haunting, sinister air. *Censored Landscapes* is a notable addition to the mounting documentation of the darkest scandal of our age." —**JM Coetzee**, author of *The Lives of Animals, Elizabeth Costello,* and *Disgrace*; winner of the Nobel Prize in Literature

"*Censored Landscapes* is a cerebral yet visceral exploration of the largely hidden world of industrial animal agriculture. La Rocca González takes an inventive approach to the subject by presenting a mosaic of narratives and photographs that engage left- and right-brain connections to farmed animals. Industry 'fact sheets' are interwoven with passages from iconic authors such as Upton Sinclair and Rainer Maria Rilke as well as the author's own narrative, including the intimate story of Henry the turkey. The trademark landscape photos of animal facilities throughout the book express a sense of dystopia, brutality, and aversion to nature. *Censored Landscapes* is an essential part of the animal liberation lexicon."—**Robert Grillo**, founder and director of Free from Harm, author of *Farm to Fable: The Fictions of Our Animal-Consuming Culture*

"*Censored Landscapes* merges poetry, clearly stated facts, lucid prose, and the stunning photography of a well-informed and dedicated scholar—a heart-wrenching and persuasive combination."—**Dr. Lisa Kemmerer** (lisakemmerer.com), professor emerita, founder of Tapestry (tapestryofpeace.org), author of *Eating Earth*

"*Censored Landscapes* is a thorough, thoughtful, and beautiful contribution to our thinking about sentience and ethics. Along with La Rocca González' prose, her photographs of quiet rural landscapes afford the viewer both time and autonomy to absorb and comprehend a new point of view of farmed animals. This is a well-crafted and revelatory book."—**Jo-Anne McArthur**, photojournalist and founder of We Animals Media

"At an early age I became aware that animal agriculture is an inherently cruel and destructive business. *Censored Landscapes* combines visual artistry, deeply personal storytelling, poignant poetry, and impactful research to reveal not only the colossal atrocity against nonhuman animals, but the resulting environmental devastation and interrelated systemic oppressions. *Censored Landscapes* is an urgently needed confrontation with the horrifying reality of farming animals."—**Moby**

"Isabella takes us on a journey that both zooms in on the poignant lives of animals freed from modern industrial agriculture and zooms out to the fortress-like exteriors of animal production facilities. The book offers a compelling personal view of what defines 'censored landscapes' that is often lonely, desolate, and complex. This unexpected coffee table masterpiece challenges you to 'not look away' but rather to analyze the unassuming architecture of today's animal agricultural landscapes that lack a sense of humanity within its barbed-wire boundaries."—**Christine Morrissey**, executive director of Harvest Home Animal Sanctuary

"*Censored Landscapes* is a powerful, thought provoking, timely text that is a must read for all. This book captures the true story of the abuse, oppression, pain, sadness, and loneliness of nonhuman animals in factory farms and other places of violence toward nonhuman animals. There are not enough books, films, and posters showing the true reality of terrorism toward nonhuman animals."—**Dr. Anthony J. Nocella II**, cofounder and executive director of the Institute for Critical Animal Studies, editor of *Peace Studies Journal*

CENSORED LANDSCAPES

THE HIDDEN REALITY OF FARMING ANIMALS

Isabella La Rocca González

Lantern Publishing & Media • Woodstock and Brooklyn, NY

2024
Lantern Publishing & Media
PO Box 1350
Woodstock, NY 12498
www.lanternpm.org

Printed in the United States of America

Library of Congress Cataloging-in-Publication Data

Names: La Rocca González, Isabella, author.
Title: Censored landscapes : the hidden reality of farming animals / Isabella La Rocca González.
Description: Woodstock : Lantern Publishing & Media, 2024. | Includes bibliographical references.
Identifiers: LCCN 2023053328 (print) | LCCN 2023053329 (ebook) | ISBN 9781590567364 (hardcover) | ISBN 9781590567371 (epub)
Subjects: LCSH: Animal welfare. | Livestock—Social aspects. | Animal industry—Moral and ethical aspects.
Classification: LCC HV4757 .L37 2024 (print) | LCC HV4757 (ebook) | DDC 179/.3—dc23/eng/20231229
LC record available at https://lccn.loc.gov/2023053328
LC ebook record available at https://lccn.loc.gov/2023053329

Contents

Acknowledgments

I would like to express my deep and sincere appreciation to my dear friends who accompanied me on photography trips and who gave me valuable feedback on my writing. Many thanks also to Harvest Home Animal Sanctuary, PreetiRang Sanctuary, Good Acres Sanctuaries, and Tribe Animal Sanctuary for allowing me to spend time with the animals in their care and for all their efforts on their behalf. Thank you also to the Culture and Animals Foundation and the Puffin Foundation for their support of this project. I'm also very grateful to Brian Normoyle at Lantern Publishing & Media, who believed in this project.

In memory of my beloved parents, Alicia Estela González and Remo La Rocca, who worked so hard to enable me to pursue my highest aspirations.

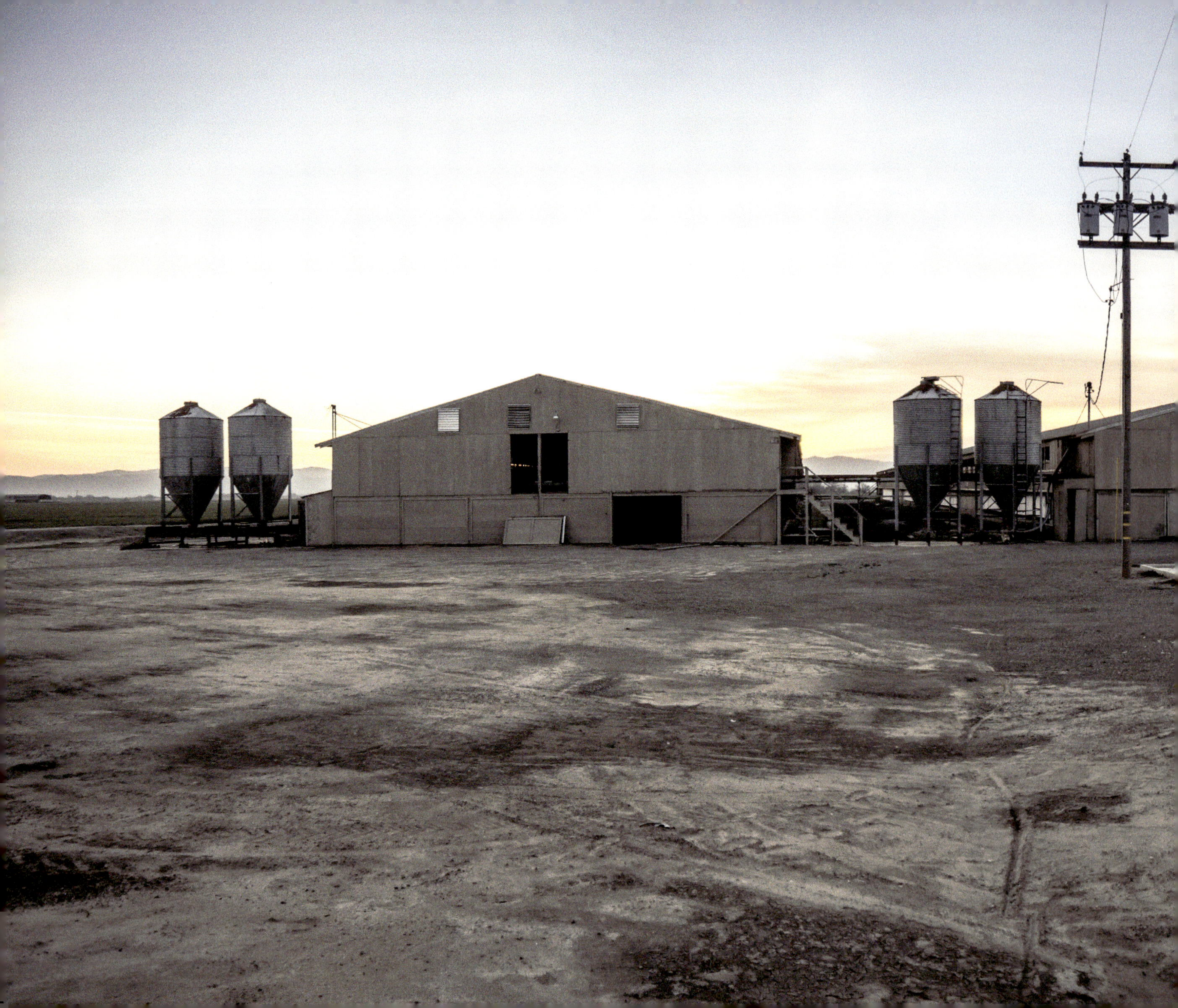

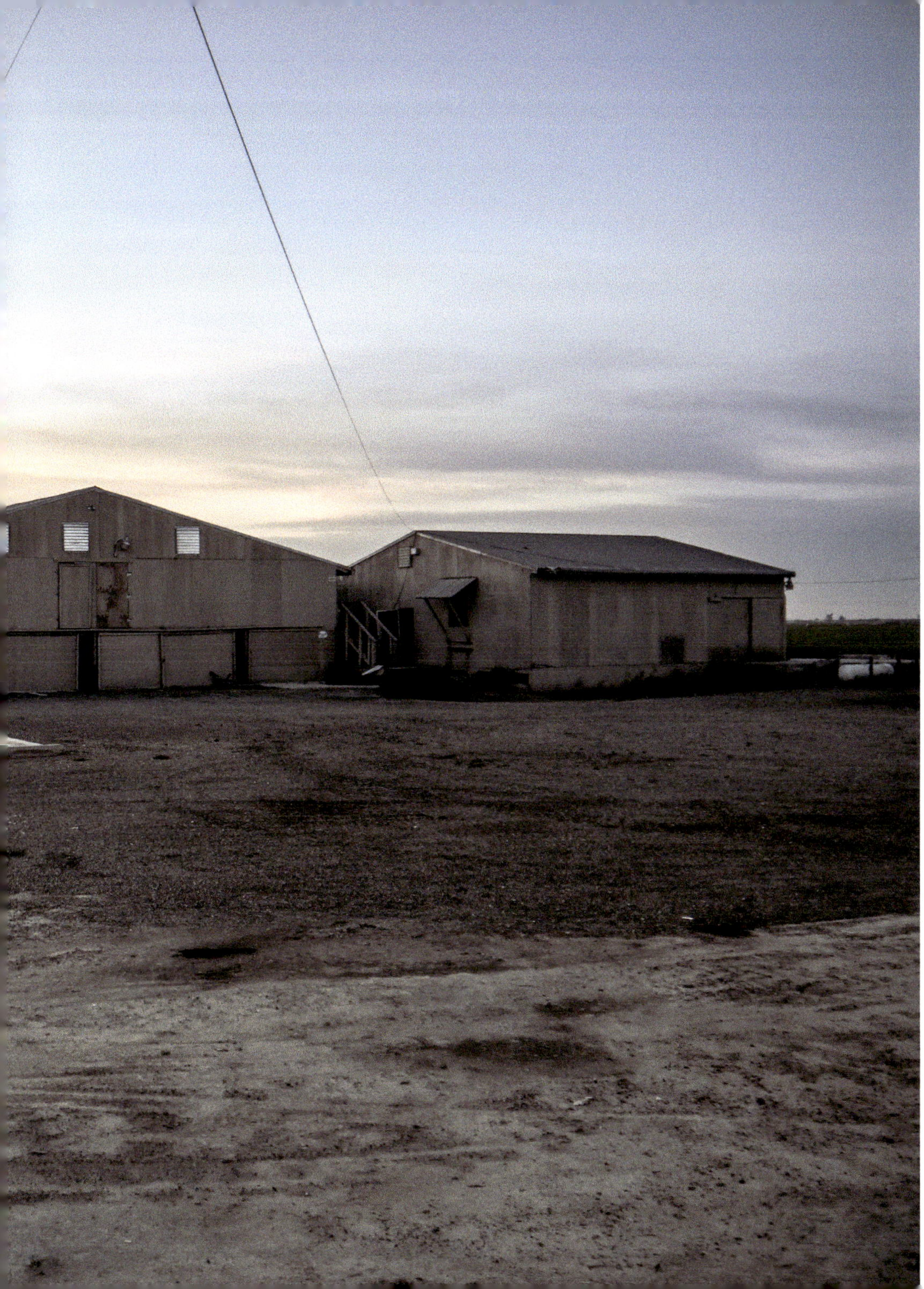

A&L Poultry
Abandoned Battery Cage Egg Farm, 2013
50,000

Inception

In 2012, in Turlock, CA, 50,000 hens were abandoned in an egg farm. Every one of those trapped birds embodied a tragic irony. Birds can fly in the sky, unassisted by metal structures or fossil fuels. They are emblematic of freedom. "Free as a bird," goes the cliché. For her whole life, each of those hens had been crammed in a rusted cage with three, four, five, even six others; her whole life eclipsed by an obsessive desire to bust out of the cage, to spread her wings and feel the sun and the rain and the breeze, to sashay on solid ground under an open sky, to make a nest and raise her babies. I imagine the grim days that followed. Now her relentless craving for escape amplified by an urgent need for food and water, the air growing louder—deafening—with the sound of 50,000 hens shrieking from thirst and starvation. Now a cagemate is dead. Now more hens immobile in the cages, their sounds muted. Now the air is thickening with the stench of rotting flesh. Second by minute by hour, two weeks passed. Finally, state officials began gassing birds by the thousands. After three more days, animal sanctuaries negotiated to take custody of the surviving 4,460 hens.

A year later, I photographed the empty facility. The filthy warehouses I found there did nothing to alleviate my nightmarish imaginings. The ammonia stink of chicken urine still lingered.

Scary things can happen to a woman with expensive photographic gear, especially when she is alone. The threat comes not only from thieves. People who breed, confine, and slaughter animals conceal their enterprise aggressively. In more than half of US state legislatures, the industry has attempted to pass "ag-gag" laws that criminalize photographing sites of animal agriculture. These laws are contrary to the First Amendment of the US Constitution but as of May 2023, they are in effect in six states.[1] The Animal Enterprise Terrorism Act is a federal law that criminalizes economic damage to an animal enterprise, including loss of profits.[2] Photographers who expose the practices of these corporations can be lumped into the same category as Timothy McVeigh and the Unabomber.

I would not have stayed long in that hellscape but for my lifelong allegiance to photography. Photography is a medium through which I make sense of the world. I photograph to translate conditions I am enchanted by, or confounded by, or disturbed by, or horrified by, into a language of light and color, and to grasp for beauty.

A photograph is a love song to the ecological implications of the landscape contained inside its frame. Carleton Watkins hauled thousands of pounds of photographic equipment through Yosemite. His photographs influenced Abraham

Lincoln to sign a bill that protected Yosemite Valley. William Henry Jackson's photographs persuaded Ulysses S. Grant to designate Yellowstone as a national park. Ansel Adams' work inspired Congress to establish Kings Canyon National Park. American landscape photography evolved in conjunction with the conservationist and environmental movement.

The tale told in photographic history books illustrates that American landscape photography, like US history writ large, has been dominated by white men. Their mythologizing of a pristine landscape colluded in the ethnocide of millions of people who lived in this country before colonization. In 1975, the *New Topographics: Photographs of a Man-Altered Landscape* exhibition was considered radical in that it depicted a stark view of human impacts on the land. Of the photographers included, Robert Adams in particular responded to the call of environmentalism. But only one of the ten artists was a woman, and none were people of color.

The tale of American landscape photography is also one of human exceptionalism. Though humans are now integral to landscape photography, farmed animals have almost entirely been excluded, despite their prodigious numbers. Terrestrial farmed animals account for 59 percent of vertebrates on earth, humans 36 percent, all other terrestrial animals only 5 percent.[3] Approximately ten billion land animals in the US are slaughtered every year.[4] The absence of farmed animals from landscape photography reflects their exclusion from environmental activism even though animal agriculture is a leading cause of climate change, deforestation, ocean acidification, habitat destruction, loss of biodiversity, and mass species extinction (see "Environmental Impacts"). This I know to be true: environmentalism and social justice must include the other animals who share our planet.

The photographs I made that February day in Turlock became the inspiration for this project. *Censored Landscapes* is a photographic exploration that tells a story in which the central characters are innocent of any crime and yet are condemned to imprisonment, torture, and death. It's a true story of ecological destruction, of worker exploitation—mostly people of color—and of secretive corporations protected by laws and enriched by government subsidies and lobbies. It is also a story that offers insight and healing.

Note: Humans are animals. In the interest of simplicity and clarity, depending on the context, I've used both "animal" and "nonhuman animal" to refer to animals who are not humans.

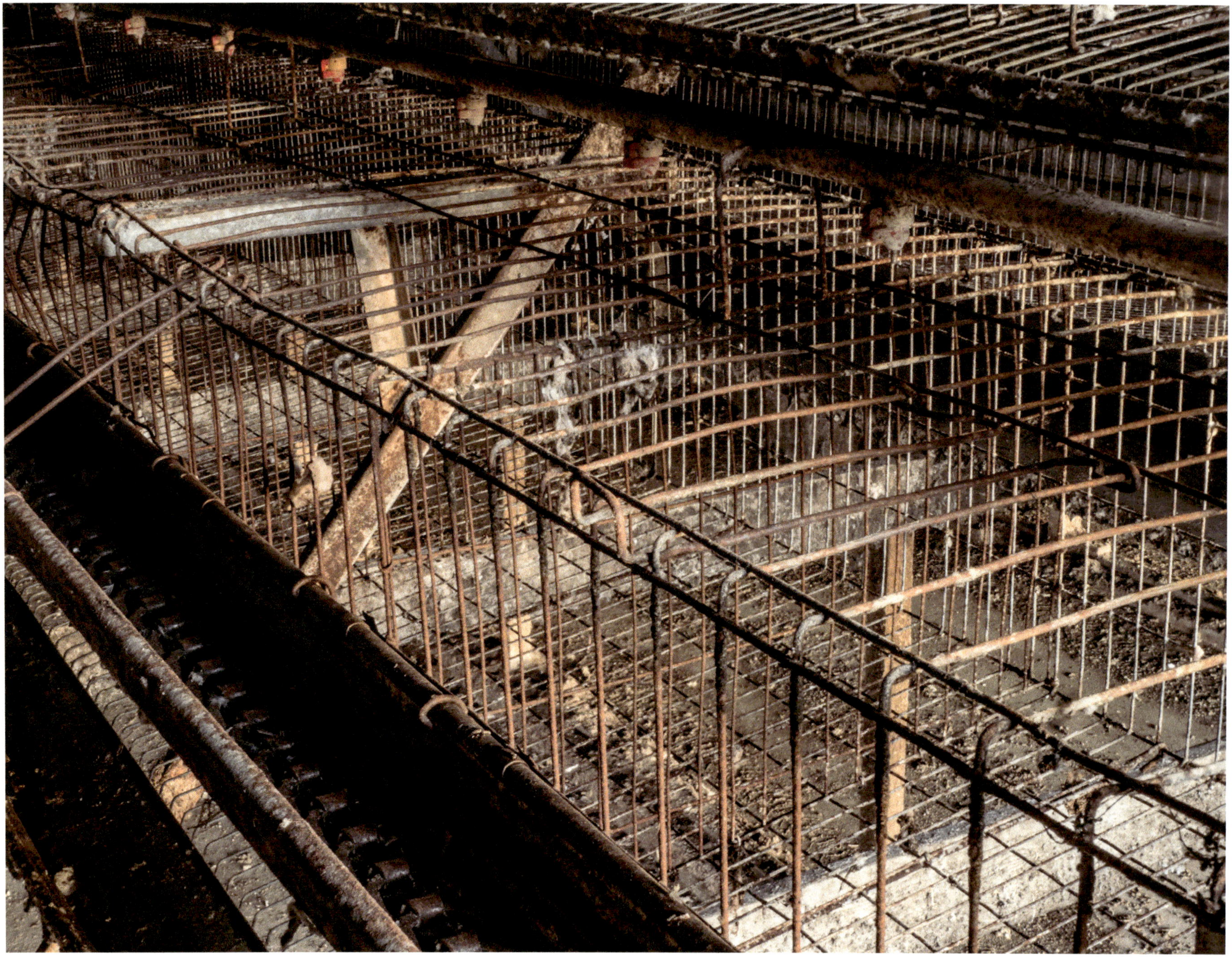

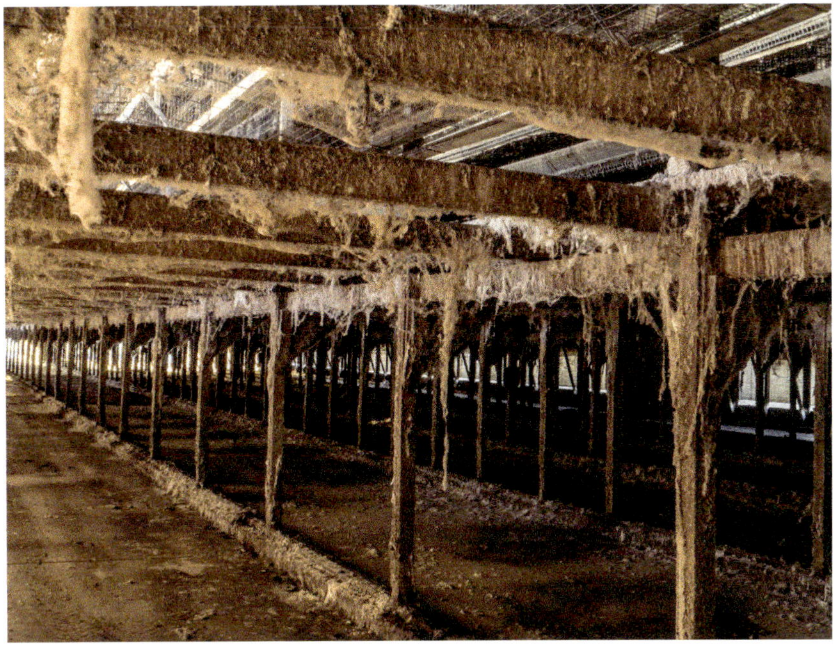

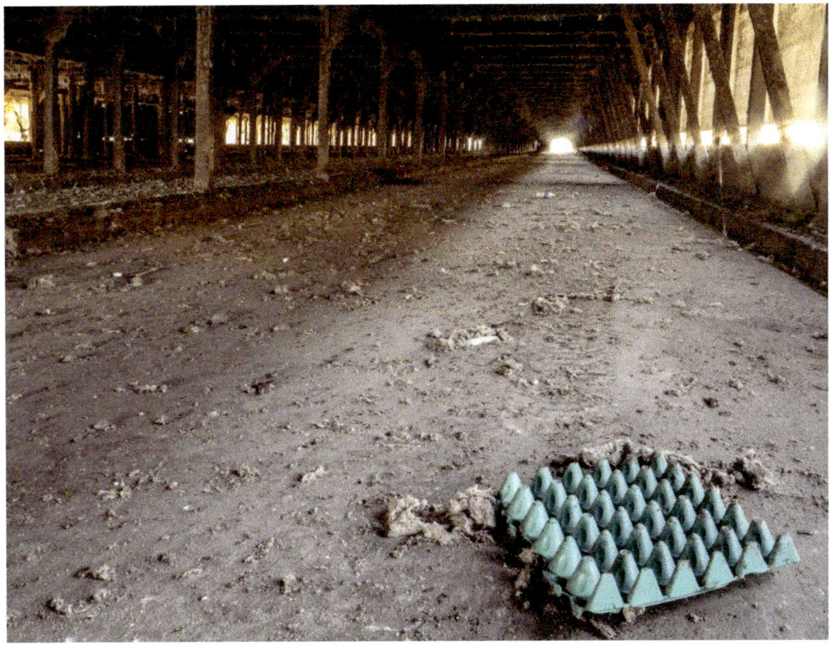

A&L Poultry

Abandoned Battery Cage Egg Farm Interior, 2013

50,000

Chickens Bred for Their Eggs FACTS

FACT: The wild cousins of domesticated chickens lay an estimated ten to thirty eggs a year.[1]

FACT: A domesticated hen, whether in a battery cage, a cage-free farm, or a backyard farm, is bred to produce hundreds of eggs during her most fertile years,[2] which causes osteoporosis[3] (often with accompanying broken bones), ovarian cancer,[4] and egg binding.[5] Peritonitis and other reproductive disorders are also common.[6]

FACT: Because they do not produce eggs, male chicks in hatcheries are killed shortly after birth[7] by gassing, asphyxiation, or, most often, grinding alive.[8] A statement released in 2016 by the United Egg Producers, which represents 90 percent of egg production in the United States, called for the elimination of male chick culling in hatcheries. As of March 2021, the practice has continued, and the United Egg Producers released a similar statement.[9]

FACTS: A hen in a battery cage facility is kept in a cage with three to seven[10] other hens. There are no legal requirements for the amount of space each hen is allotted, but most hens confined in battery cages have about sixty-seven square inches each of metal cage each to stand on for their entire lives.[11] A letter-sized piece of paper measures ninety-three and a half square inches. Hens kept in battery cages cannot dust bathe, walk, forage, stretch their wings, stretch, nest, or roost. They can sustain injuries from the cage wire and from overgrown claws.[12]

FACT: In 2021, over 70 percent of eggs were produced in battery cages.[13]

FACTS: USDA consumer packages labeled "cage-free" indicate that the hens are not kept in battery cages. They "vary from farm to farm, and can include multitier aviaries." Hens in these facilities are not required to have access to the outdoors, nor are there any requirements as to the amount of space each hen is allowed.[14] Not all eggs labeled cage-free are graded by the USDA—"USDA cage-free" labeling is a voluntary fee-based service.[15] For cage-free facilities to remain profitable, the hens are most often intensively confined. Cage-free facilities tend to have increased levels of ammonia, dust, and bacteria, which can cause eye infections and respiratory problems.[16] [See *Organic, Cage Free, Humane Certified Egg Farm (Chickens), 2015.*]

FACT: Hens that lay eggs with a USDA "free range" label must have outdoor access for 51 percent of their lives. However, producers are granted the label based only on their own description—there are no on-farm inspections to verify. There are no outdoor space requirements and no requirement that the animals spend time outdoors.[17]

FACT: The label "pasture-raised" is not federally regulated for egg-producing hens and has no enforceable meaning.[18]

FACT: The beaks of baby chicks in the egg industry, whether battery-cage or cage-free facilities, are seared off with no anesthesia.[19]

FACT: A chicken's beak is more sensitive than a human hand.[20]

FACT: Caged hens in the egg industry are starved for approximately ten days at a time until they lose 25–35 percent of their body weight to stimulate egg laying.[21] As of 2013, 80 percent of caged hens were starved.[22]

FACT: A hen in the egg industry is slaughtered when she is two to three years old, when she no longer produces enough eggs to be profitable.[23] (For a description of the slaughtering process, see "Bad Boss.")

FACT: A chicken's average lifespan is six to eight years, though some may live up to thirty years.[24]

FACT: Chickens are excluded from the United States Federal Humane Methods of Slaughter Act[25] and from the Animal Welfare Act.[26]

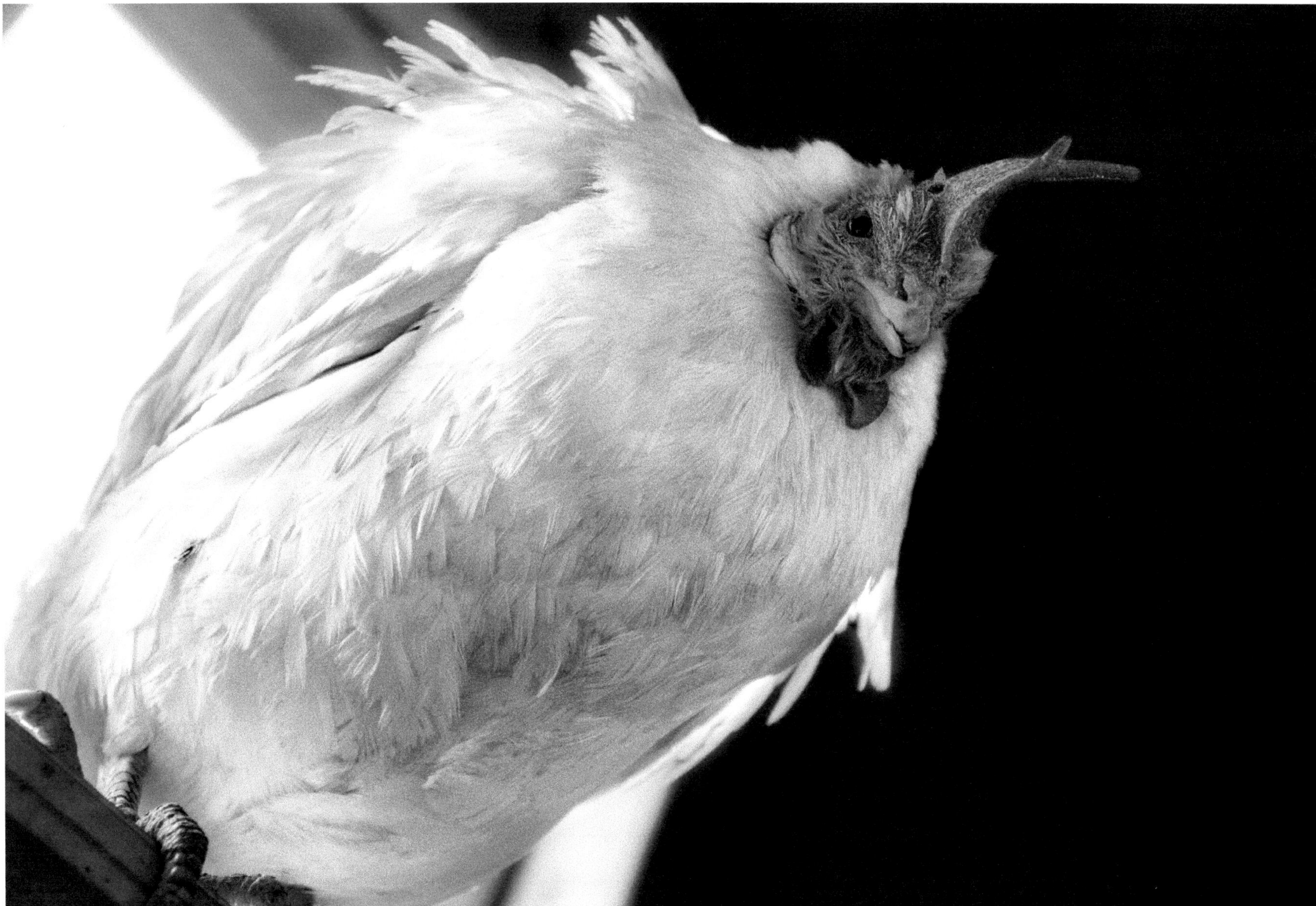

Kelsey—A&L Poultry Egg Farm Rescue, 2015

Connection

I've named it the Dreadful Dreaded Dread. Like seaweed that's always present but mostly appears at low tide, the DDD emerges at 2:28 a.m. Or 3:47 a.m. Or 4:13 a.m. No one else I know is awake. There's no one to call. Full moon or new moon, starry sky or starless, it doesn't matter.

I try to slow my breath, but it rushes ahead. The more control I attempt, the less tranquil I am. *Practice equanimity,* I tell myself. *Detach.* But no. It creeps in my guts, in my joints, in my synapses, in the crazed beating of my heart. The DDD will not be lulled or dismissed. It is splendidly amorphous and ingeniously antsy.

The infrastructure for the DDD is in my DNA. My father was nine years old when he was orphaned in World War II. His family had been grievously poor; the war flung him and his siblings into greater poverty. More trauma ensued. When I was growing up, my dad suffered night terrors that ripped his sleep apart. I'd awaken to him yelling in the other room, terrified. Epigenetic intergenerational trauma: trauma that leaves a mark on one's genes. Future generations inherit it. I likely inherited it.

Like black flies on sticky tape curling down from the ceiling, all sorts of worries adhere to the DDD. A partial list, in no particular order: bed bugs; incontinence; bioterrorist attacks; dementia; homelessness; incontinence and dementia while homeless; infestations of invasive insects; plastics, plastic in the ocean, microplastics; tornadoes; fossil fuels, global warming; monocultures; dying pollinators; endangered wildlife; animal agriculture; environmental collapse. And then the one that transcends even death: terror of being all alone in an indifferent universe. Out there beyond Earth's atmosphere swirls a mess of gasses and masses and particles and black holes. In that unfathomable and impersonal vacuum, our most sublime experiences and values—beauty, joy, kindness, empathy, altruism, love—are mere abstractions. What if our noblest ideals are nothing but social constructs, mere human invention that feeds a delusion that life is not meaningless or futile, just human strutting and fretting, sound and fury in a tale told by an idiot, signifying nothing? *

I suppose for some people, religion provides reassurance. I have piled my plate high from an all-you-can-eat buffet of religious beliefs and practices. What I'm left with is that I do not have the kind of faith required for most religions. I cannot take it on faith that the qualities I hold dearest exist beyond human constructs, that we are not alone. I need evidence.

The morning light tends to wither the Dreadful Dreaded Dread. Sunshine and emotion are linked; sunlight increases serotonin levels.[1] Light refracts into rainbows and suffuses the world with color. Under optimum conditions, humans see a range of light from violet to red.[2] Bees, for example, and butterflies also perceive ultraviolet light.[3] The visible light spectrum for chickens includes all the colors of human vision plus ultraviolet and infrared.[4]

But sunlight is not efficient in the production of eggs for human consumption. In the warehouses where hens like Kelsey are imprisoned, artificial light is alternated with periods of darkness to stimulate maximum egg laying.[5] This for an animal whose perception of light is richer than that of humans.

Chickens and humans are much more the same than we are different. Chickens and humans share more than half our genes, a genome more elemental than the epigenetic trauma I may have inherited.[6] Sentience—the capacity to experience feelings—connects humans with birds, with fishes, with mammals. Chickens recognize beauty,[7] protect their young, and exhibit empathy. A rooster signals danger from predators with various calls depending on the threat. If he finds a yummy morsel, he broadcasts to others.[8] Anyone who has seen baby goats together cannot doubt they are jumping with joy. Stressed fish seek out caresses even from robot fish.[9]

The cow bellows and is beaten back as her calf is dragged away bleating. Is this not love? Humans are not isolated in our own little love bubble. I've known this since I was a kid. Most kids know this. I would gaze into the magnificently lashed chocolate eyes of the horse who grazed in a field near my house. We'd meet up at the fence anytime I walked by. She loved when I stroked her coat and cooed at her. I loved when she nuzzled me and blew at me from her nostrils. When we were together, oxytocin, the love hormone in mammals,[10] surged for both of us. When falling in love, when playing with the dog or cuddling with the cat, when a mom breastfeeds her baby, oxytocin is doing its thing. In other animals, oxytocin has its counterpart: isotocin in fishes, mesotocin in birds and reptiles.[11]

The existential fear of being all alone in an indifferent universe is an illusion, a squalid cage erected by the human mind. We, all animals, are made from stars.[12] We are of the earth and the earth is of the universe, a truth so absolute as to be trite. The line drawn between human and nonhuman is make-believe. Every sentient animal on earth is connected by DNA and stardust and goodness and love, evidence that we are not alone.

* *Macbeth* Act 5, Scene 5, 19–28
"Tomorrow and tomorrow and tomorrow
Creeps in this petty pace from day to day
To the last syllable of recorded time,
And all our yesterdays have lighted fools
The way to dusty death. Out, out, brief candle!
Life's but a walking shadow, a poor player
That struts and frets his hour upon the stage
And then is heard no more. It is a tale
Told by an idiot, full of sound and fury,
Signifying nothing."
—William Shakespeare

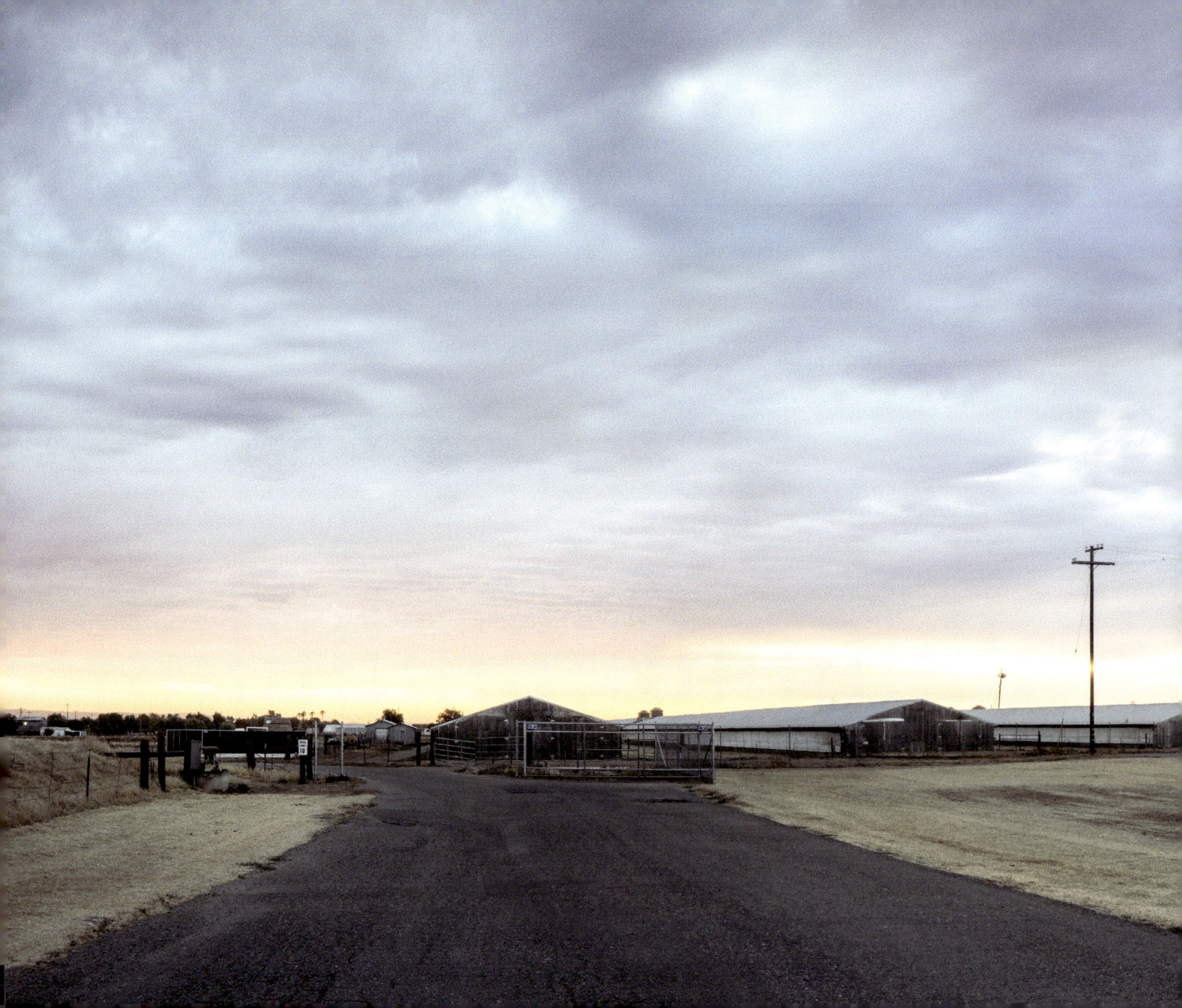

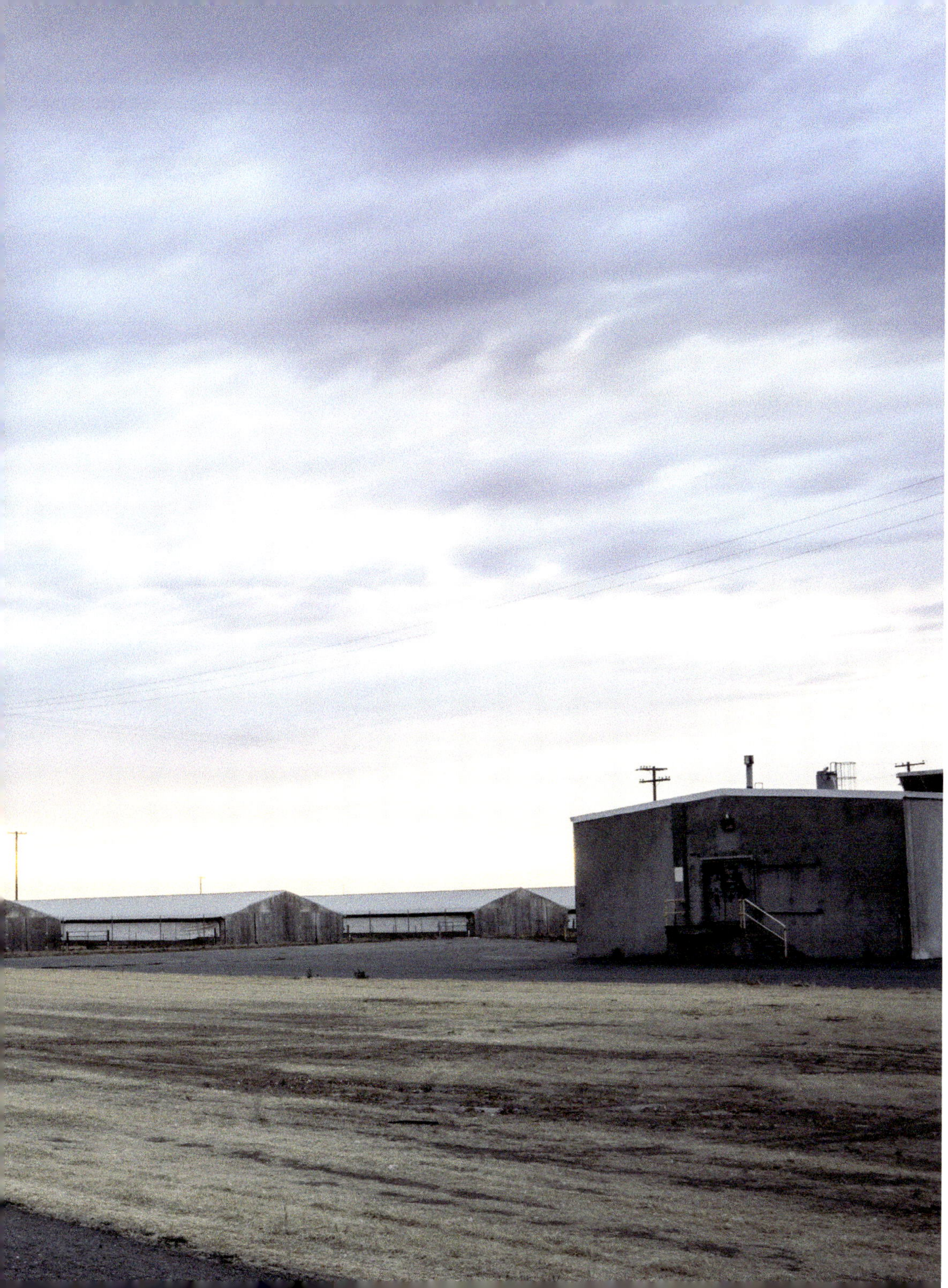

Poultry Farm (Chickens and Turkeys), 2014
349,867

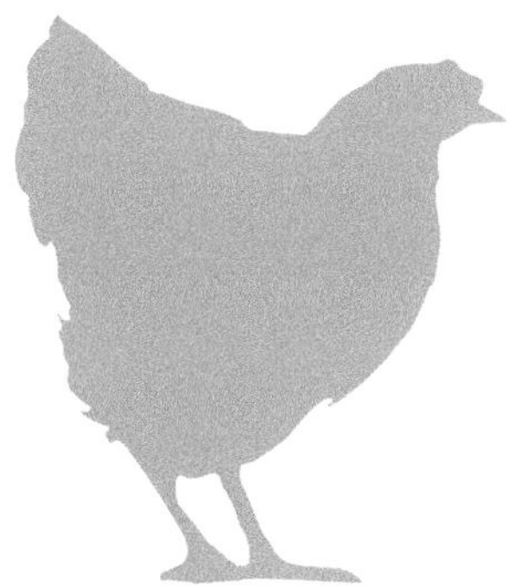

Numbers and Names

In the viscous August heat of a Southern big box store parking lot, I thought I heard a mewling baby. Preparing to shatter a window, I followed the sound to a late model pickup truck. The windows were cracked open. There was nobody inside. Underneath the truck, a kitten cowered, meowing his big-eyed, big-eared little head off.

I couldn't leave him, but I didn't think I should take him home to my other cats. I've never cared for more than two cats at a time, a feat of rigorous self-control especially in the face of the millions of adorable felines in desperate need of homes. I feared that the sluices would break and next thing you know, three cats would become nine, would become a hoarding situation. But no one in the store claimed him and rescue organizations had no place for him. I took him home. No regrets. He made our happy home even happier. And I've managed to keep the number of cat roommates to three, albeit with some difficulty.

All three of my cats are individuals with distinct personalities, habits, and preferences. They are as different from each other as humans, and as similar. People who live with companion animals understand this.

In the agricultural industry, nonhuman animals do not exist as individuals but as capital to be exploited for profit. The word "capital" comes from the Latin *capitalis* meaning "head," in the sense of a head of cattle. In medieval Europe, the number of cattle owned by a family stood for wealth and prestige. Cattle were used as currency for large transactions.[1] In the US, mercenary corporations are granted personhood, meaning they are entitled to at least some of the legal status, rights, and protections of humans. They can own land and money.[2] Nonhuman animals, however, are deemed property, with none of those rights or protections.

How then to portray the individuals hidden away in remote places and windowless warehouses?

The numbers included with these photographs represent the number of animals bred, confined, or slaughtered in the landscape. Though the word "number" is different in origin than the word "numb," here there's a relationship. The enormity of the numbers draws a numbing veil over the individuality of each animal. The numbers are revealing, but they're a pitiful substitute for even one of the animals whose identity, sentience, emotions, and existence is obliterated.

Accessing the numbers wasn't easy. They've mostly been provided by government agencies after extensive and tedious research. In some cases, the number of animals can only be calculated; they exist as "animal units," a perplexing term based on a thousand-pound cow.[3] In one case, the number is "proprietary," meaning the public is not privy to the information (see *Livestock Auction, 2016*). Other businesses were not required to report the data, in which case I asked an industry expert for an estimate.

Except for the first farm (see *A&L Poultry Abandoned Battery Cage Egg Farm, 2013*), the owners of which were prosecuted as a matter of public record, I have not advertised the names or locations of the operations I photographed. This project does not focus on the kind of graphic depictions of cruelty found all over the Internet, of men beating turkeys with iron rods and throwing them aside or slamming piglets onto a concrete floor by the hindlegs. Atrocities against animals are rife in the industry, but I don't have the heart to investigate. Most heartbreaking is the innocence of the animals. They can't possibly understand why this is happening to them—why the contempt, why the cyclonic rage. Beyond unbearable pain, they must feel utterly forsaken by all things good. The men must also feel bereft of goodness, to be so possessed by monstrous resentment or whatever drives them to viciously harm those creatures. My intention is not to expose any specific worker, company, or owner. The dingy, wretched structures revealed in those videos insinuate everyone and everything—farmed animals, wildlife, humans, rivers, oceans, forests, soil, air—into their abyss. *Censored Landscapes* scrutinizes animal agricultural industry standards, business as usual.

Chickens Bred for Their Meat FACTS

FACT: The majority of chickens bred for their meat are called "broiler chickens."[1]

FACT: Increasingly, chickens bred for their meat are artificially inseminated. (A description of the process can be found in "Nonconsensual Sex.")[2]

FACT: Chickens bred for their meat have been genetically manipulated to grow breasts and thighs so enormous they often cannot hold up their own weight.[3]

FACT: Chickens bred for their meat spend most of their short lives standing in the feces and urine that collects in the litter on the shed floors.[4]

FACT: Broiler chickens are slaughtered when they are seven to nine weeks old.[5]

FACT: A chicken's average lifespan is six to eight years, though some may live to be thirty years old.[6]

FACT: More than nine billion "broiler" chickens were slaughtered in the US in 2021.[7] (A description of the slaughtering process can be found in "Bad Boss.")

FACT: Chickens are excluded from the United States Federal Humane Methods of Slaughter Act[8] and from the Animal Welfare Act.[9]

FACT: The United States Department of Agriculture allows companies to slaughter chickens at a rate of 140 birds per minute, though some facilities have received waivers that allow them to slaughter the birds at 175 per minute.[10]

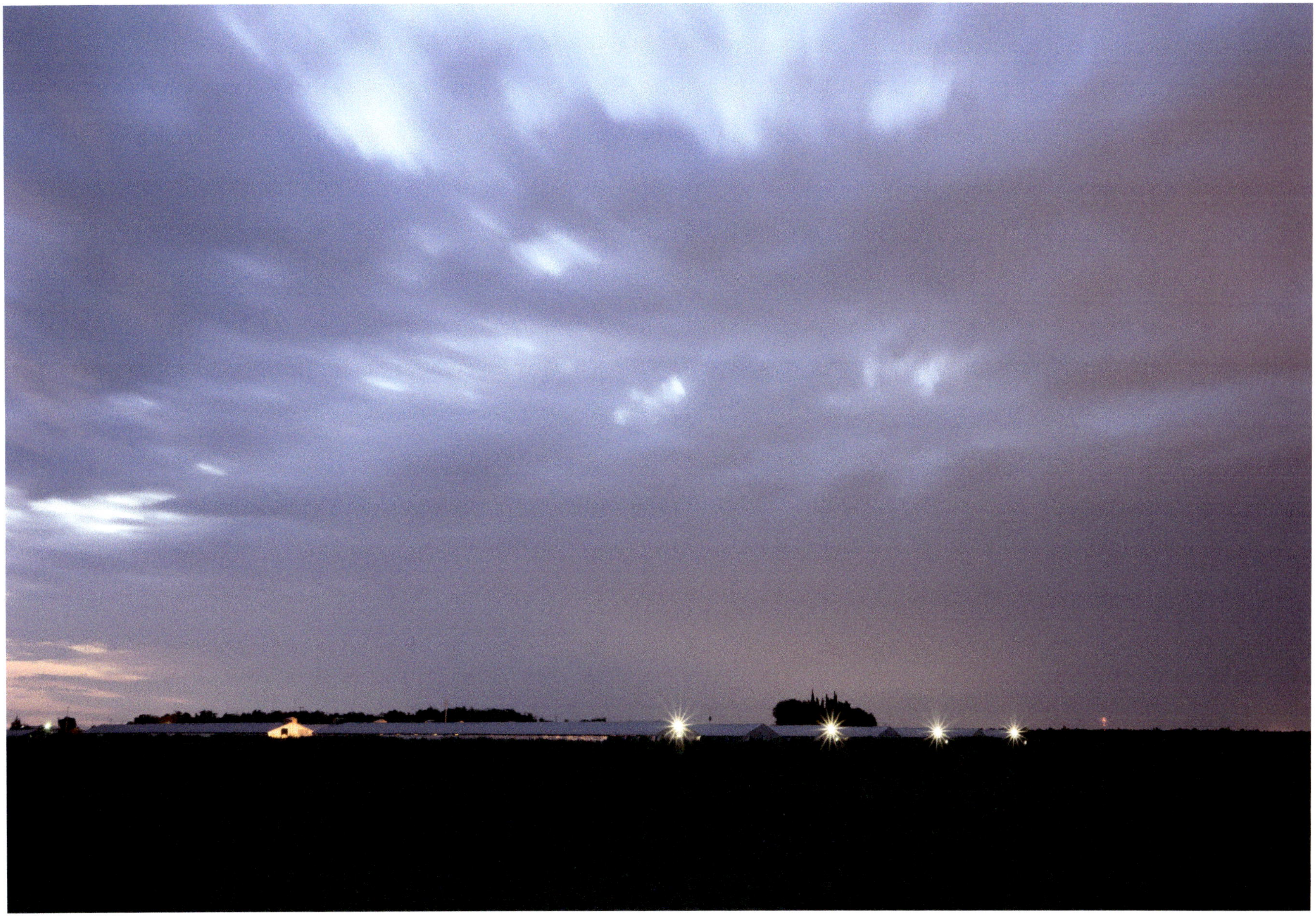

Chicken Farm, 2014
609,280

Foie Gras FACTS

FACT: Foie gras (French for "fat liver") refers to the diseased, enlarged liver of a male duck or goose after force feeding (gavage).[1]

FACT: The process of force feeding involves inserting a six-to-ten-inch metal tube in the throat and into the stomach of a duck who is about three months old. A mash of corn and fat weighing up to a pound (about 20–30 percent of the duck's weight) is forced into the duck's stomach with a mechanical, pneumatic, or hydraulic device two to three times a day for twelve to fifteen days.[2]

FACT: Ducks raised for foie gras exhibit aversion to the person who force feeds them.[3]

FACT: Force feeding causes pain and injury to the duck's esophagus.[4] The force-fed ducks will pant and produce semiliquid feces.

FACT: At the end of force feeding, the duck's liver is up to seven to ten times its normal size. [5] The duck's enlarged liver often results in the inability of the duck to move or stand.[6]

FACT: The great majority of ducks raised for food are intensively confined in crowded cages in which the ducks cannot stand erect, turn around, or flap their wings.[7]

FACT: Ducks' natural habitats are near water, such as lakes, marshes, and rivers. Natural waterfowl behavior includes bathing and swimming.[8] Wild ducks can fly long distances at speeds up to sixty miles an hour.[9]

FACT: Ducks raised for foie gras are slaughtered within two weeks of force feeding at less than four months old.[10] (A description of the slaughtering process can be found in "Bad Boss.")

FACT: A duck's natural life span averages five to ten years,[11] and they can live up to twenty-nine years.[12]

FACT: Ducks are excluded from the United States Federal Humane Methods of Slaughter Act[13] and from the Animal Welfare Act.[14]

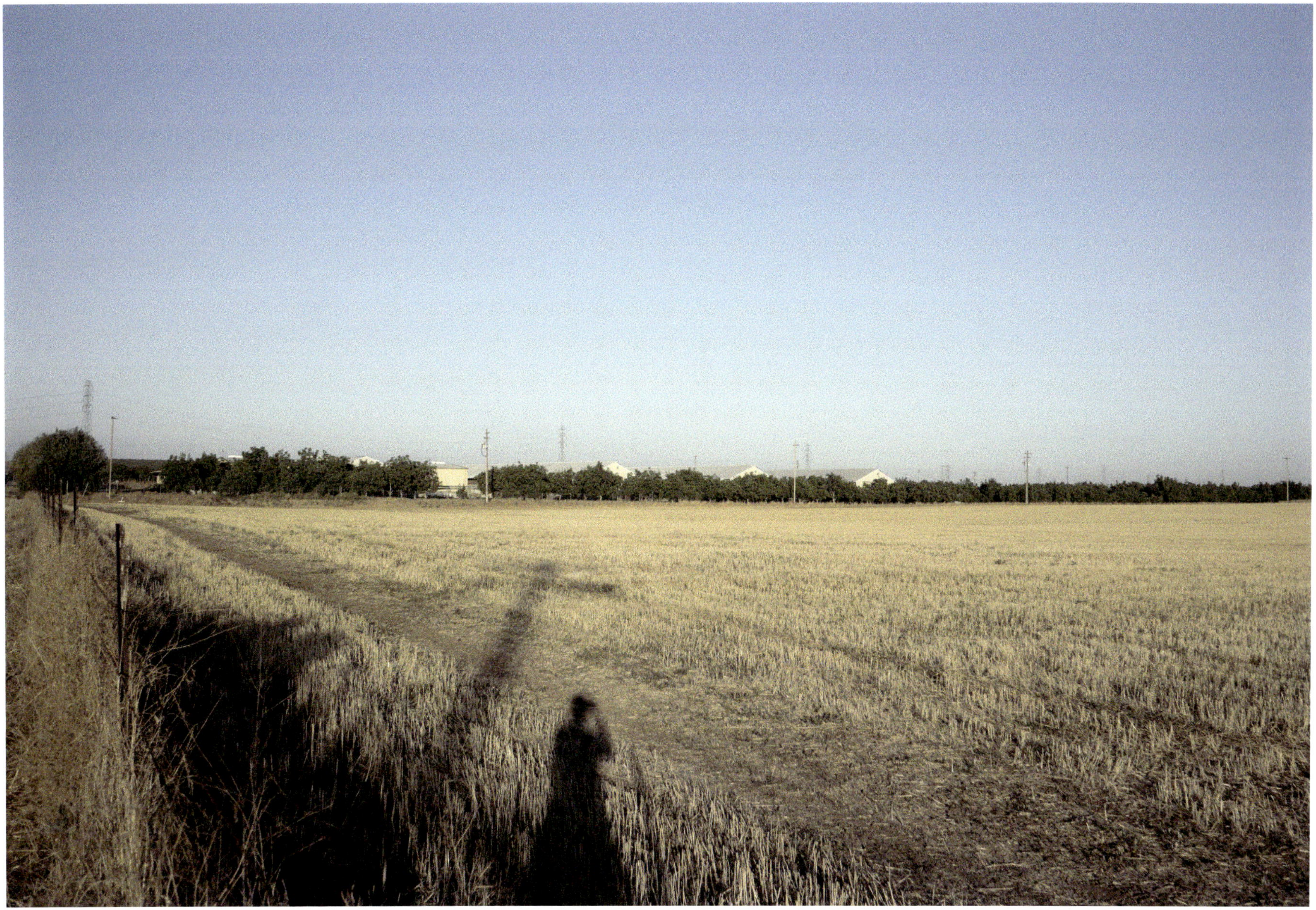

Egg Farm (Chickens) and Former Foie Gras Facility (Ducks), 2014
300,000

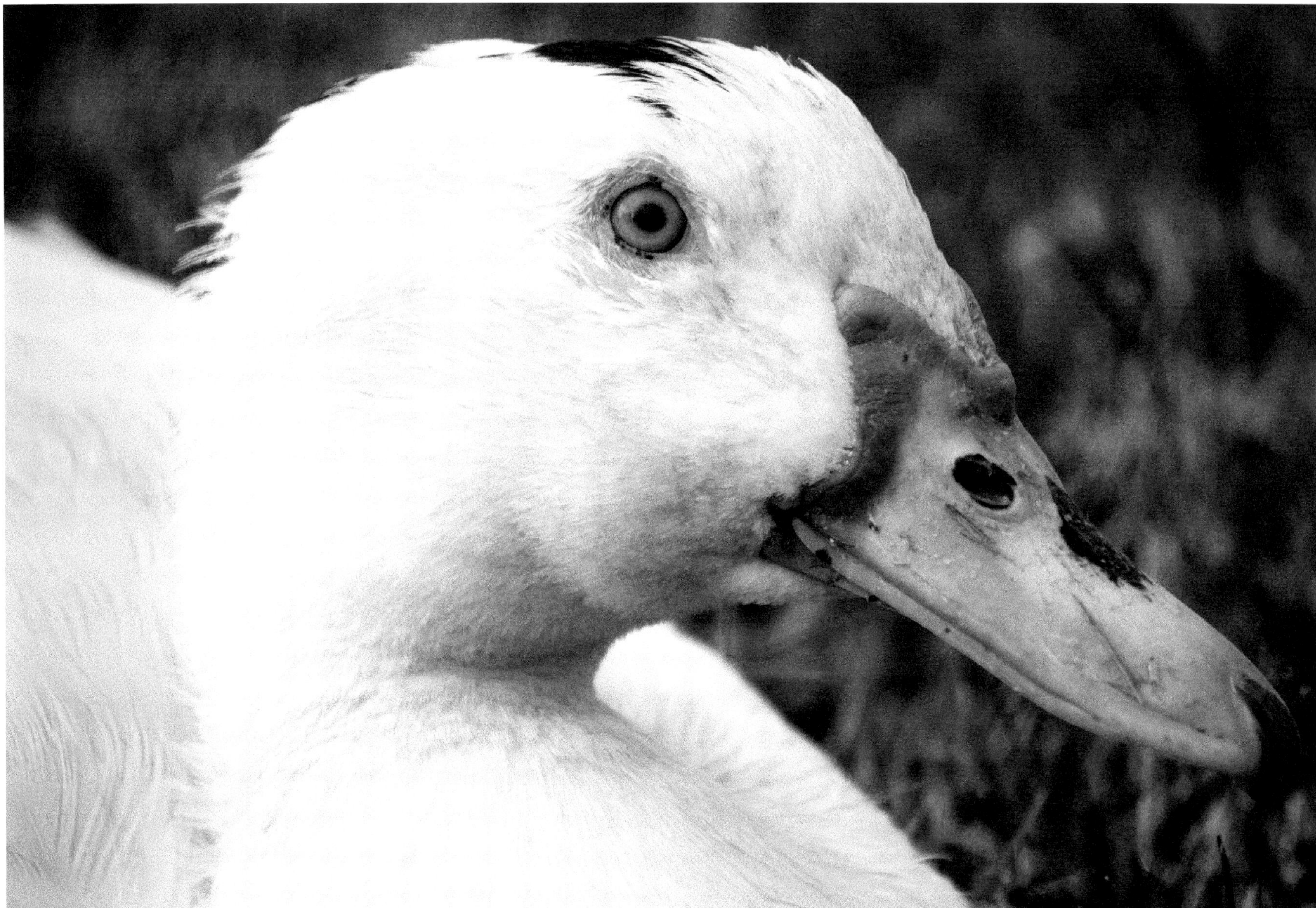

Bentley—Foie Gras Rescue, 2014

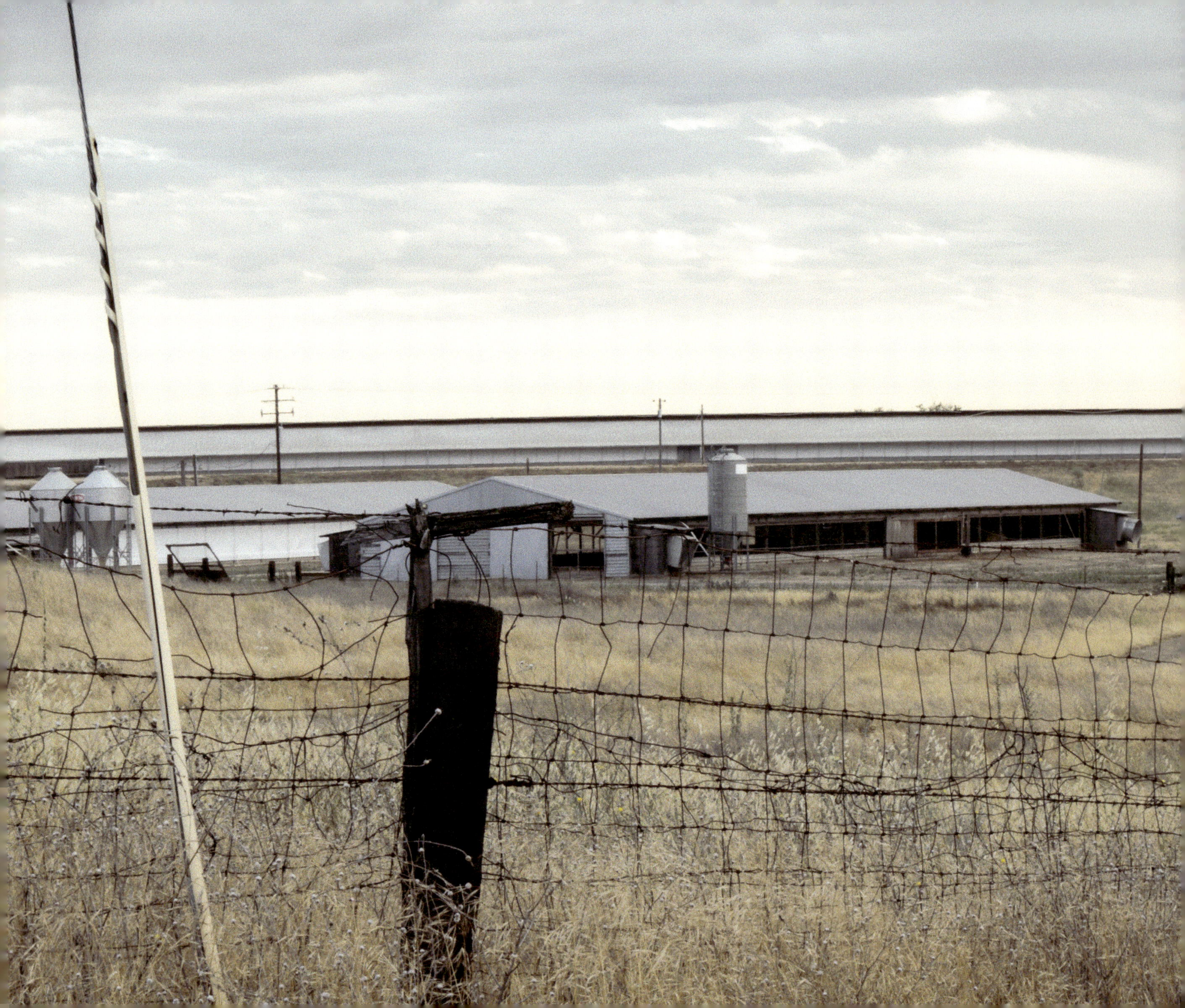

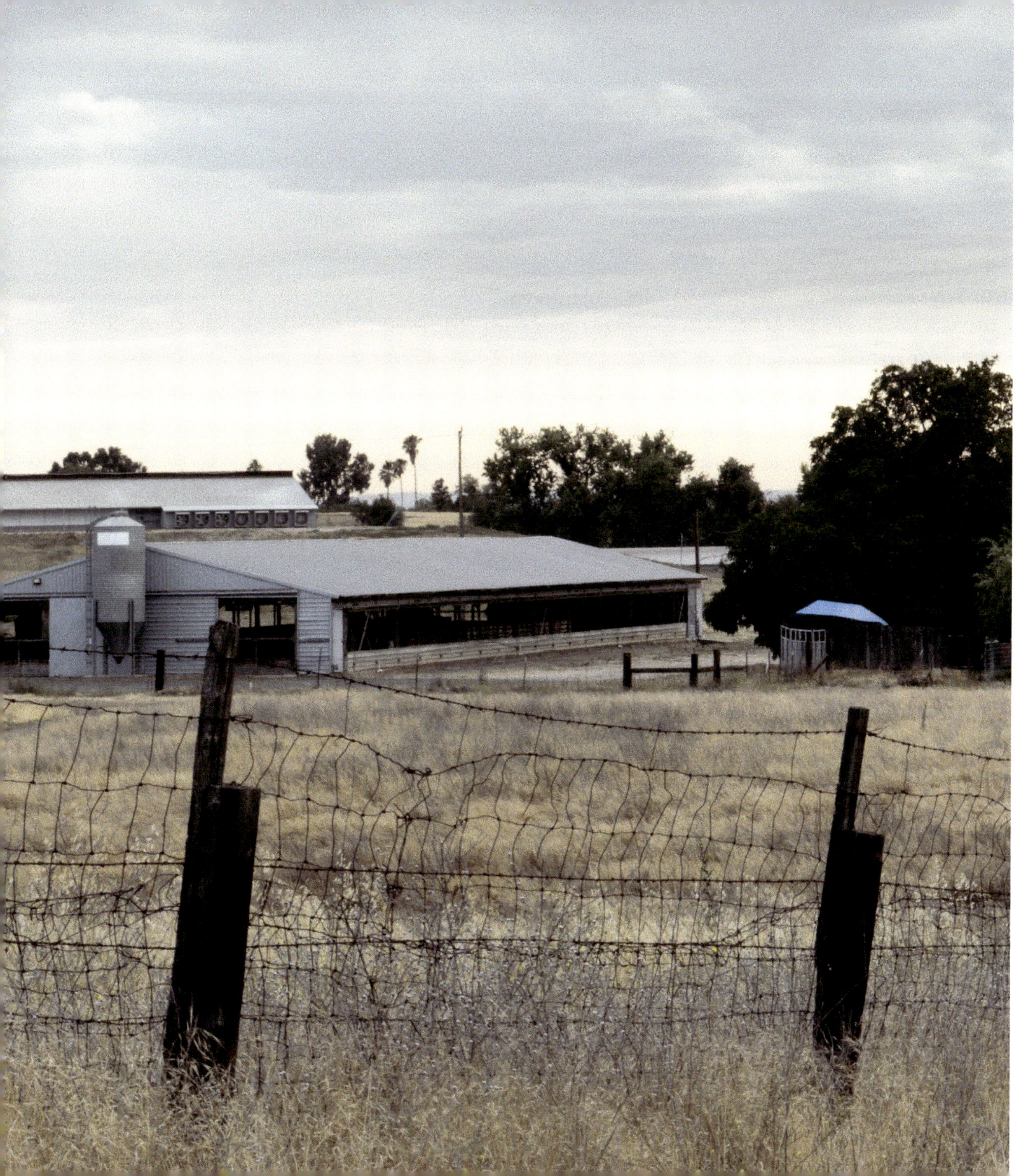

Turkey Farm, 2014
37,666

FARMED VERSUS WILD TURKEYS FACTS

FACTS:

- Over 200 million farmed turkeys are slaughtered for food in the US each year.[1] In 2022, 208.2 million federally inspected turkeys were slaughtered for food.[2]
- In 2022, there were an estimated five million wild turkeys in the US.[3] Before colonialist logging, clear cutting, and hunting, there were an estimated ten million wild turkeys in the US.[4] The population of wild turkeys continues to decline.[5]

FACTS:

- Farmed turkey hens are bred to grow to about fifteen and a half pounds within their first two months of life.[6]
- Wild turkey hens reach a mature weight of eight to twelve pounds[7] at about one to two years of age.[8]

FACTS:

- Male farmed turkeys are bred to grow to about thirty-eight pounds at four to five months of age.[9]
- Male wild turkeys reach a mature weight of about fifteen to twenty-five pounds[10] at about seven months of age.[11]

FACTS:

- An estimated 99.8 percent of farmed turkeys are confined on industrialized farms[12] where each animal has about two and a half to four feet of space.[13]
- The home territory of a wild turkey can range from 400 to 2,000 acres or more[14] and may include fields, marshes, pastures, and forests.[15]

FACTS:

- Farmed turkeys are unable to fly due to their unnaturally large size and intensive confinement.
- Wild turkeys can fly at average of fifty-five miles per hour and run up to twenty-five miles per hour. They are also able to swim.[16]

FACTS:

- Farmed turkeys are fed corn and soybean meal[17] in addition to antibiotics (in 2018, almost 1.5 million pounds of antibiotics were fed to turkeys).[18]
- The diet of wild turkeys includes seeds, nuts, acorns, buds, leaves, roots, tubers, fruit, fern fronds, insects, and salamanders.[19]

FACTS:

- Male farmed turkeys are slaughtered at about four to five months then sold as cutlets, sausage, franks, and deli meats. Female farmed turkeys are slaughtered at about two and a half months and usually sold as whole birds.[20]
- The average lifespan of a wild turkey is 1.3 to 1.6 years and they can live as long as thirteen years.[21]

FACTS:

- Because of their grotesque proportions, it is not cost effective for farmed turkeys to mate naturally, therefore they are artificially inseminated.[22] (A description of the process can be found in "Nonconsensual Sex.")
- In a springtime mating ritual, a male wild turkey vocalizes to attract a female. Once he has her attention, he struts and dances with his snood and wattle bright red, regally fanning out his tail feathers and puffing out his iridescent body feathers, until the female turkey consents.[23]

FACTS:

- Farmed turkey hens are bred to lay eighty to one hundred eggs in twenty-five weeks, after which most are slaughtered.[24]
- Wild turkey hens lay one clutch—about four to seventeen eggs—each year.[25]

FACT:

- Turkeys are excluded from the United States Federal Humane Methods of Slaughter Act[26] and from the Animal Welfare Act.[27]

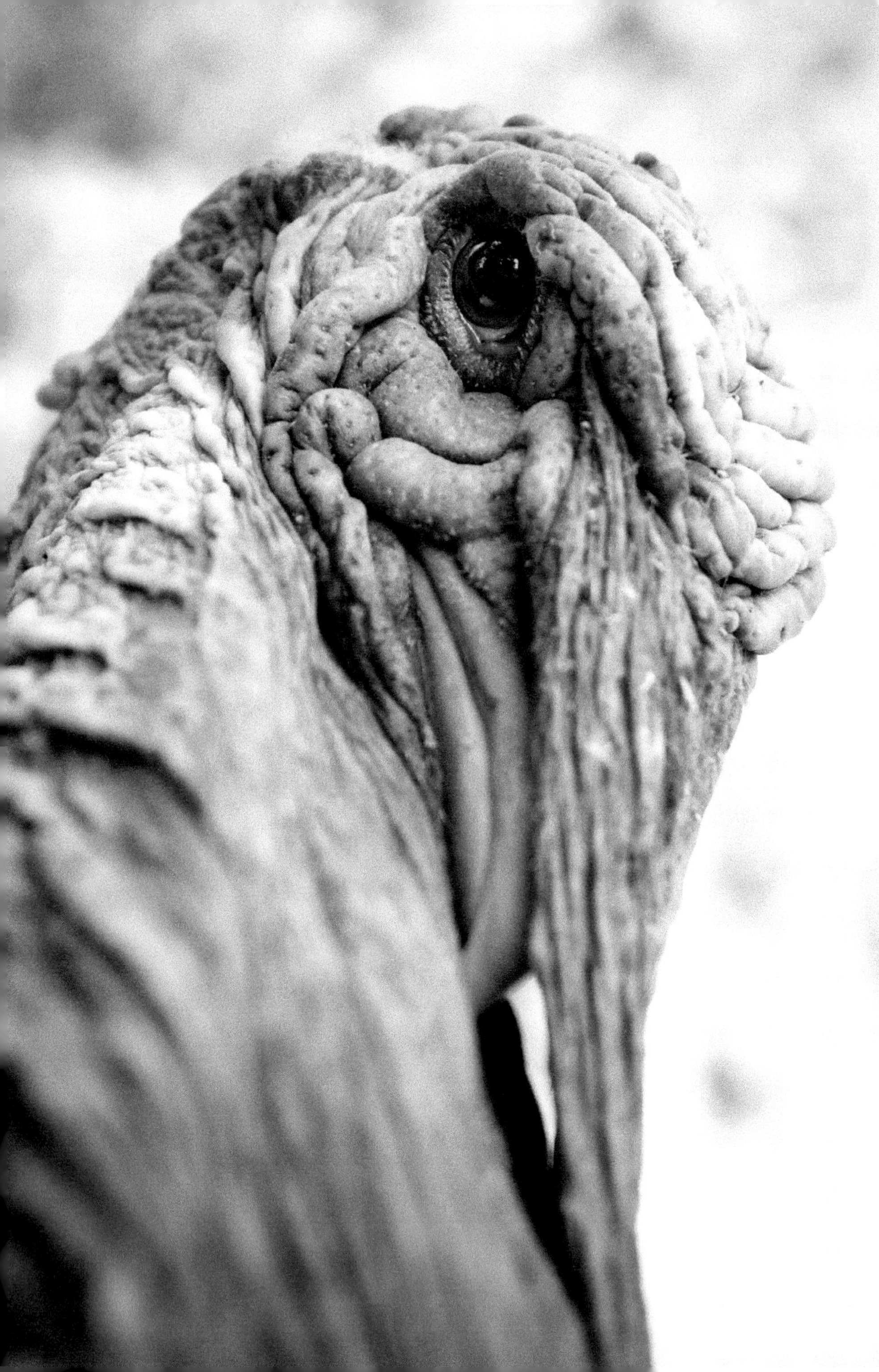

Henry—Turkey Farm Rescue, 2015

Photographing Henry

Example One: Outside a convenience store, a skinny woman with a ruined, exquisite face leaned against the wall. She intermittently sucked from a straw sticking up from the top of a Big Gulp. A street photographer aimed her lens at the woman and released the shutter, click click click.

"What the hell you think you're doing?" The woman was now too close for focus. She popped the lid off the Big Gulp and tossed the contents of the drink at the photographer's crotch.

Example Two: Inside a Tesla Roadster, a movie star quarreled with his fashion model girlfriend. The car glided to a stop next to the curb in front of a parking valet. An aspiring young paparazzo pointed his lens at the model, click click click, then at the movie star as he strode around to the passenger side, click click click. Anger bloomed on the movie star's face. He grabbed the model's gold-chained shoulder bag and began swinging at the paparazzo. The bag grazed the photographer's head and he scuttled away.

Photographing humans without their permission is a risky practice while, as a rule, attempts to photograph nonhuman animals are rarely met with aggression. Flight rather than fight tends to be their response. Wildlife photographers are regarded as honorable; they "shoot with a camera, not a gun," proclaims the meme. Farmed animal owners and their minions make for the exception to the rule.

Example Three: A row of cypress trees obscured the pig barn. I parked the car about four hundred feet away. My friend and I debated whether walking on the dirt road that wound behind the trees would be trespassing; no signs were posted. We decided that my friend could act as lookout and I would venture down the road with my camera. If anyone came, she would slam the car door, our signal to make myself scarce.

I had hardly set foot on the dirt road when a truck pulled up behind my car. A corpulent white man stomped out of the truck toward me, blustering about me taking pictures and threatening to call the police. My friend invited him to please do so; we weren't doing anything illegal, and he was verbally assaulting us. I don't recall the bellicose man's exact words, just that we needed to get the hell out of there. My friend and I climbed into my car and locked the doors. For the next twenty-some miles of our trip, the truck followed close behind until I managed to lose him by turning right then left after a traffic light that was yellow for me, red for him.

I photographed many of the animals depicted in this project, including Henry, in the safety of animal sanctuaries. People who spend little time thinking about animals in terms other than their use to humans may not be aware that increasing numbers of farmed animal sanctuaries have sprung up in recent years. These are not zoos. An animal sanctuary's only purpose is the happiness and well-being of the animals who live there. They are marvelous places.

The day I met Henry, I resisted photographing him. I hung out on the ground with my camera around my neck. Henry moved with deliberate grace, as if carrying inside him something weighty and precious. Generations of animal agriculture breeding had transformed the bronze iridescence of his wild ancestors to a shimmering white that, if Henry had been slaughtered as originally intended, would have produced flesh more appealing to consumers.

Henry promenaded over to me, studying me with wary, soulful eyes. I greeted him in a voice gentle as warm mist. I spoke to him of the weather, of the trees, and the birdsong, of his beauty. He edged nearer and nearer, listening closely. I sang a Mexican *canción ranchera* my mom used to sing. He pressed against my arm. He liked when I caressed his head. He especially liked when I scratched under his wings. Then he placed his dinosaur foot on my sneaker. This became his customary greeting.

In the 1994 book *Looking at Photographs: Animals*, A. D. Coleman writes, "A portrait is a picture that suggests something about that particular individual's personality. Animals have identities, habits, moods, emotions, and expressions just as humans do, and photographers often find ways of revealing those characteristics."[1]

I don't disagree with the statement as far as it goes, but it leaves out so much. Portraiture implies a tacit agreement among humans based on a set of conventions that have developed over thousands of years. A portrait is meant to memorialize the sitter. If the photographer is an artist, the portrait becomes part of an oeuvre. Coleman's definition also discounts complicity between the subject and the portraitist. The human sitter gives the photographer permission to interpret their identity for other humans.

The ways animals recognize themselves or others have little to do with silver gelatin, ink, or dye on a flat substrate—or with an array of tiny red, green, and blue lights emitted from a screen. These meager impressions barely hint at the richness and breadth of animal perceptions, or at the movements and smells and sounds and light and presence in all dimensions that characterize an individual. My animal subjects can't be bothered with my memorializing, or my interpretations, much less my oeuvre. They are much more concerned with my being.

Humans confuse a flat abstraction with the thing itself. People point to their phone screens and say things like, "This is my new motorcycle" or "That's me and my boyfriend in Costa Rica" or "This is my dog." Studies show that some animals recognize humans from two-dimensional images; somehow, scientists have inferred that human reductionism is evidence of superior cognition.[2]

René Magritte pointed out the absurdity of this in his painting, *La Trahison des images (The Treachery of Images)*; below an image of a pipe the text reads *C'eci n'est pas une pipe* (This is not a pipe). The thing humans identify as a pipe is not a pipe at all, it is oily pigment dried on canvas.

I've had mixed feelings about photographs of animals. In the 1980 book *About Looking*, John Berger observed that "the commercial diffusion of animal imagery began as animals started to be withdrawn from daily life. . . The reproduction of animals in images—as their biological reproduction in birth becomes a rarer and rarer sight—was competitively forced to make animals ever more exotic and remote."[3]

Animals in photographs are often mere compositional picture elements, humorous or aestheticized objects, means of advancing human careers or influence. They are seen as entertainment, targets of ridicule, consumer goods, inferior and alien beings, almost always colored by the entrenched speciesism that infects human society. Why would animals in my photographs be seen differently?

The animals I photographed for this project serve as ambassadors of their species. Their images are meant to individualize the vast numbers of farmed animals bred, confined, and slaughtered. The thing I've most treasured about making their portraits is that, like thoughtful conversation or dancing tango or singing harmony, portraiture presents a compelling opportunity to intertwine my consciousness with another's. I delight in the process.

Rest in peace Henry, my cherished friend.

Supermoon

Rosemary Alles

You
walked to me through
the hope
of that
silvered night.
The night of a Supermoon,
a night that was
to be
my last.

My world and yours
touched that night.
You, standing in freedom
and I
huddled amongst hundreds
waiting
waiting to die.

Through the burnished moment
my flesh shivered.
Imagining
you would cut
through
rusted
metal
so I could walk
as you did.
Bathed in the white
light
of a Supermoon.

I was four years old.
Used.
Milked.
Made pregnant
again,
and
again.
My children
taken,
gone.
I am worthless.
Spent
and ready
to die.

"Dairy Cow"
they called me.
But.
You knew
I had
another name.

So you came.
That night.
You came
to call me.
Once.
By my real name.
Earthling.
Sister.
Creature of the stars.
Waiting.
Waiting to die.

We stood.
You and I,
barely touching.
Your world
and mine
divided by iron
united by a moon.
A full moon
that was as close
to earth
as it would
ever be.
My last moon
a Supermoon.

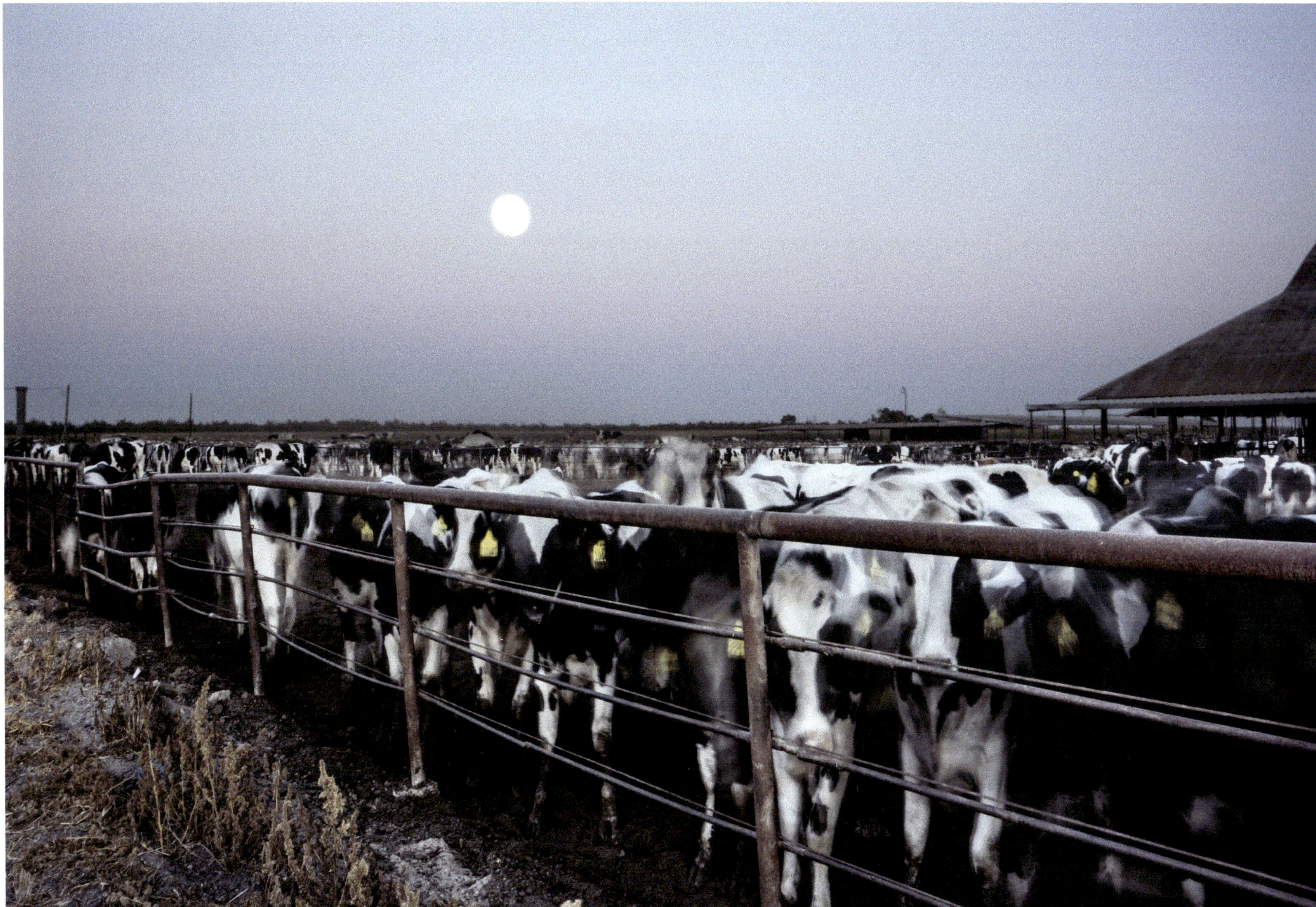

Night of the Supermoon Dairy Farm, 2014

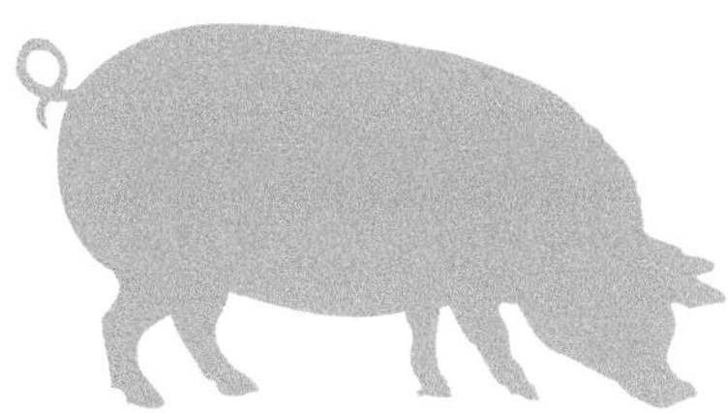

Bad Boss

In 1974, forty men were each paid four dollars and fifty cents—the equivalent of about thirty dollars in 2023—just for showing up for a scientific study about learning and memory. Each participant was instructed to administer electrical shocks of increasing intensity every time a learner responded incorrectly to a question that tested his memory. A supervisor pressured the participant to continue to shock the learner even after the learner cried out in pain and warned about his heart condition. Though their payment was guaranteed, all but one of the participants followed the boss's orders and continued to administer shocks. Twenty-six of the participants continued to shock the learner until they reached the maximum of 450 volts (a severe and dangerous shock).

The participants were unaware that the learner was an actor and that he received no shocks.[1] The experiment has been criticized as unethical and its methodology discredited, yet it continues to provoke debate about why people comply when the boss's orders are inhumane, abhorrent, or evil.[2]

I am more intrigued by the one participant who refused to cooperate with the boss. What kind of individual defies authority just because it's the compassionate thing to do? More compelling is the person who does so at a greater cost to himself. Virgil Butler was such a man. Virgil began working in chicken processing plants at the age of fourteen and continued, on and off, until he was thirty-seven years old. After nine years employed at a Tyson slaughterhouse, where he participated in killing as many as 80,000 chickens in a shift, Virgil left the industry and became an animal rights activist.[3] Of his experience he wrote:

> You begin to feel a sense of disgust at yourself at what you have done and continue to do. You are ashamed to tell others what you do at night while they are asleep in their beds. You are a killer.
>
> People tend to avoid you, even others at the plant, whether from instinct or because they know what you do and can't understand how you can do it night after night. There must be something wrong with you. You have the smell of death on you. You are a killer. A mass murderer.
>
> You shut down all emotions eventually. You just can't care about anything. Because if you care about something, it opens the gate to all those bad feelings that you can't afford to feel and still do your job. You have bills to pay.[4]

I'm not implying that slaughterhouse workers are in any position to defy authority. Workers like Virgil are typical of workers in the animal slaughtering and processing industry; they are from rural communities, without resources for higher education, short of opportunity, and poorly paid.[5] More than half are people of color (34.9 percent Hispanic, 22.5 percent Black).[6] The prevalence of Perpetration-Induced Traumatic Stress, a form of Post-Traumatic Stress Disorder (PTSD)[7] and Serious Psychological Distress (SPD), is higher among slaughterhouse workers than among the general US population.[8] Slaughterhouse work has been connected with higher rates of drug and alcohol abuse.[9] Research in communities where slaughterhouses are located demonstrates higher rates of violent crimes, property crimes, and drug offenses.[10]

Workers in the US meat industry suffer an estimated three times as many serious injuries as the average American worker, including repetitive stress injuries, fractures, burns, head trauma, and amputations.[11] Respiratory disorders are rampant.[12] The rate of injuries and illness is likely higher than estimated; many workers refrain from reporting for fear of retribution. Employers may also underreport to avoid higher costs.[13] Workers do not receive proper safety training and disproportionately lack health insurance.[14]

During the COVID-19 pandemic, the moving average of COVID-19 cases in meat-packing-dependent counties in the last two weeks of April 2020 was more than ten times that in other rural counties. The average was nearly seven times the number of daily cases seen in other rural counties by the end of May 2020.[15]

In 2022, 9.54 billion chickens, 208.2 million turkeys, and 26.7 million ducks were slaughtered for food in the US.[16] Poultry industry guidelines recommend withholding food and water from the birds for eight to twelve hours before they are slaughtered. The reality is that the time between feeding and slaughter may last much longer than recommended; after feeding, the birds are captured, packed into crates, loaded onto trucks, transported, unloaded, and then made to wait for their turn to be slaughtered.[17] Companies are not required to report time in transit, nor are there legal limits on the duration of transport or wait time at the slaughterhouse. There are no requirements to protect the animals from the elements or provide ventilation during transit.[18] One study documented transport to take as long as 12.8 hours.[19] Each of the phases involved in arrival at the slaughter facilities causes more suffering for the birds.[20] About twenty million chickens die on the way to slaughter every year.[21]

At the slaughterhouse, the birds are hung by their legs in metal shackles, which causes them more stress, pain, and injury.[22] They are carried upside down to an electrified bath meant to stun them, their necks are sliced, and they are dunked in scalding water.[23] The speed at which the USDA allows facilities to slaughter birds continues to increase. In 2023, the rate is 140 birds per minute though some facilities have received waivers that allow them to slaughter 175 birds per minute.[24] The stunning technique can be ineffective, so many of the birds are conscious as the blade cuts their throats.[25] Or the saw misses and they are boiled alive. Almost one million chickens and turkeys are boiled alive every year in slaughtering facilities.[26] Birds raised for food are excluded from the US Federal Humane Methods of Slaughter Act[27] and from the Animal Welfare Act.[28]

Larger animals such as cattle, pigs, sheep, and goats also suffer multiple stressors on their way to the slaughterhouse; the only federal regulation meant to protect these animals in transport is the Twenty-Eight Hour Law, which requires that animals transported in vehicles and vessels should be unloaded for food, water, and rest after twenty-eight hours with some exceptions. This law does not protect birds.[29] Enforcement of even this minimal requirement is doubtful.[30] At the slaughterhouse, the animals are goaded through a chute to the kill area with electrical prods. Political scientist Timothy Pachirat, who worked undercover at a Great Plains cattle slaughterhouse, wrote,

> Gilberto and Camilo use the electric cattle prods extensively, sometimes sticking them under the animals' tails and into their anuses. The cattle jump and kick when shocked

> in this way, and many also bellow sharply. Gilberto uses the prod in almost rote fashion, shocking practically every animal, especially as they near the hole in the slaughterhouse wall that leads into the knocking box. Even when the cattle are tightly packed, with the nose of one animal pushed up against the rear of the animal in front of it—sometimes even with its head squished between the hind legs of the animal in front of it—Gilberto still delivers the electrical shock, often causing the cow to mount the animal in front of it. Already caked in feces from their time in the feedlot, the transport truck, and the slaughterhouse holding pens, the cattle are packed so closely together as they push their way up the chutes that the defecation of one animal often smears the head of the animal immediately behind it.[31]

The "knocking box" refers to the area where the animals are stunned. The most common stunning method is a blow to the animal's skull with a captive bolt gun.[32] The average cattle "knocker"—the employee who fires the bolt gun—stuns around 2,500 animals a day, or one every twelve seconds.[33] Then the animals are shackled by their hind leg, hung upside down, and stuck with a knife in the neck or chest to exsanguinate or bleed them out.[34] Citations for stunning violations at federally inspected plants rose to 38 percent by 2015. At state-inspected plants, citations for ineffective stunning rose to 46 percent.[35] In 2019, the US Department of Agriculture rolled out a New Swine Inspection System (NSIS), which requires only voluntary inspection and removes speed limits in the slaughter of pigs.[36]

Homo sapiens have existed for at least 300,000 years[37] and have only been farming animals less than 12,000 years,[38] a bizarre blip on this end of the human timeline. North and Central Americans did not farm animals until the invasion of the European colonists about 400 years ago.[39] Veggies, legumes, grains, nuts, and fruit are the basis for superb cuisines of all flavors. Plant-based restaurants of every variety are proliferating like wildflowers in spring. Meat, dairy, fish, and egg alternatives are available in all chain grocery stores and even at fast food restaurants. But no matter, Americans continue to consume more animals than ever.[40] Animal agriculture is a bad, ruthless, avaricious boss whose demands result in ecological devastation (see "Environmental Impacts"), grievous health effects (see "Eating Animals and Human Health"), and colossal suffering.

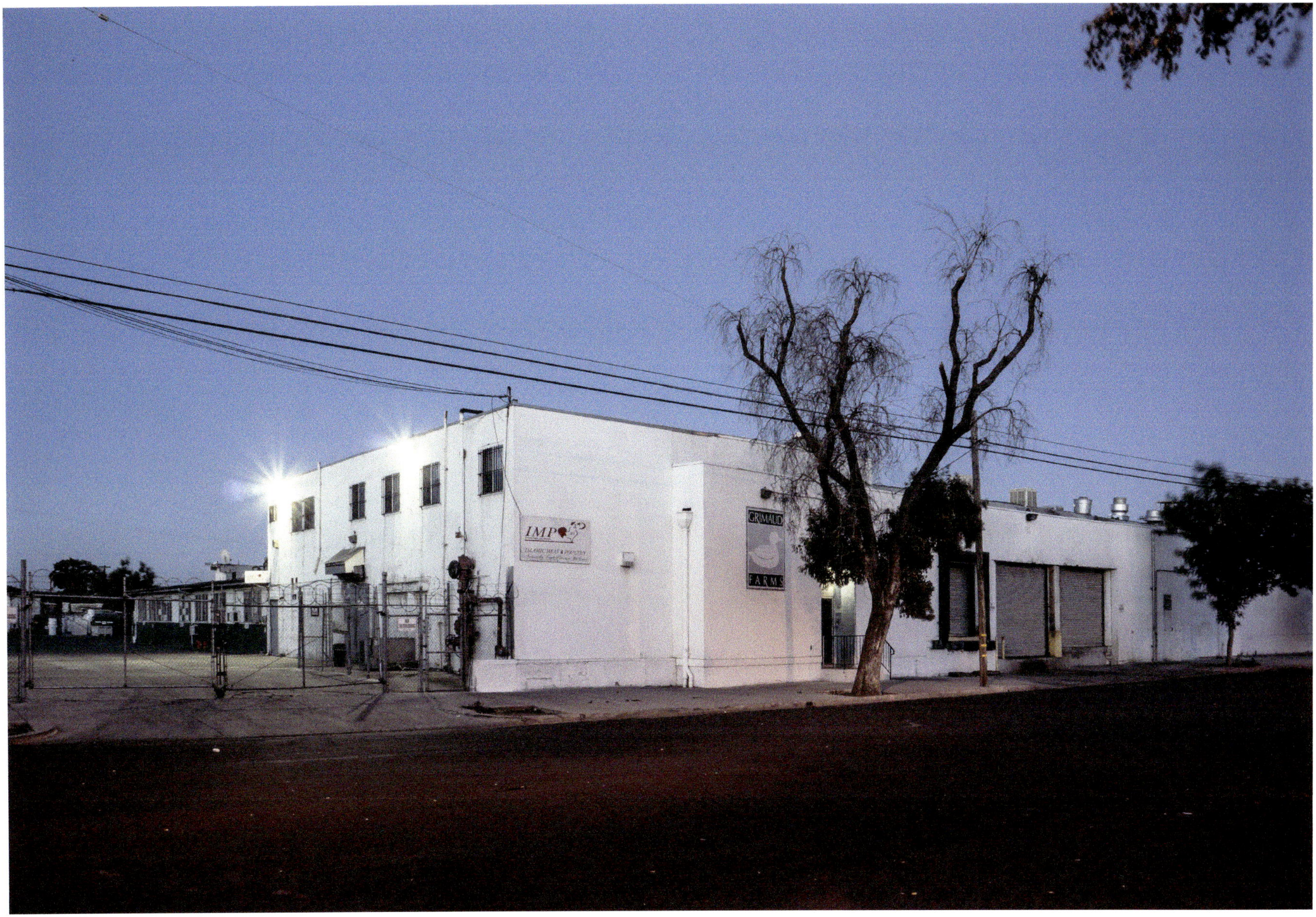

Slaughterhouse (Chickens, Ducks, Guinea Fowl), 2015
1,300,000–3,640,000/year

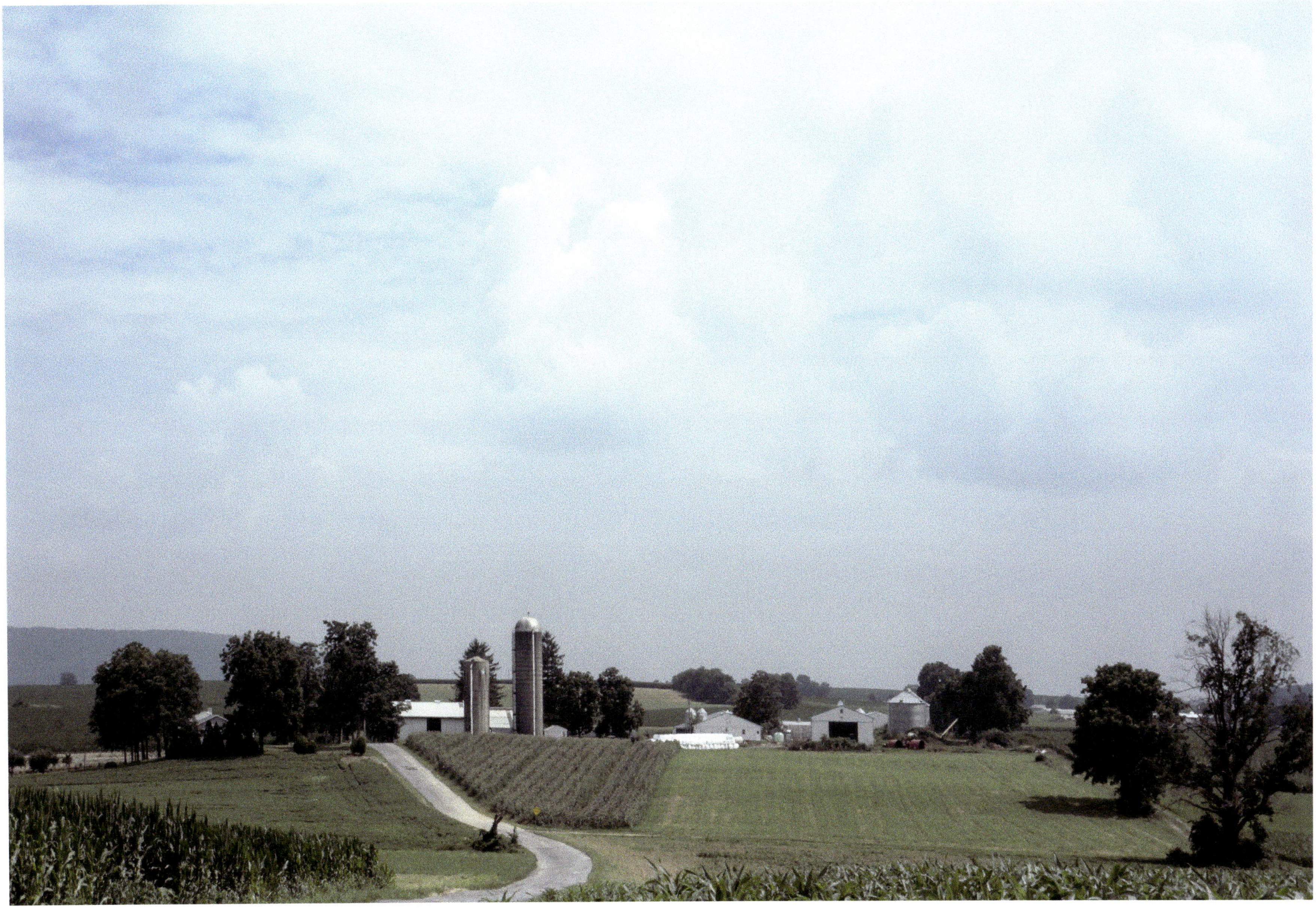

Pig Farm, 2015
8,900

A Landscape Rewritten

Mona Seymour

The Critical Animal Studies literature has of late illuminated for me the intersections of violence toward farmed animals and toward farm and slaughterhouse workers. I will probably never again think of the pain, suffering, fear, and distress experienced by sentient nonhuman animals raised for profit and consumption without also calling to mind the unpleasant, dangerous work environments and exploitive work conditions endured by largely undocumented immigrant laborers in these industries. This landscape, like all of those pictured in this collection, embodies concealed violence and invisibilized interlocked oppressions of farmed animals and farm workers. It stood out to me, though, for the possibilities that it also evinces. The possibility here is for a landscape rewritten in the image of veganic agriculture.

Veganic agriculture is an approach to organic plant agriculture that builds soil fertility using plant and rock-based inputs, such as green manures, plant-based composts and mulches, plant meals, seaweed, and rock phosphate, while avoiding animal-based inputs, such as manures and blood and bone meals. It also evinces respect for local wildlife communities, often seeking to integrate them into the holding; landscape elements such as native flowers, hedgerows, and ponds attract pollinators and natural predators. Organic farming techniques such as crop rotation and intercropping abate disease and insect pressure. Concern for safe and fair working conditions is also often embedded in veganic farming principles. A veganic system thus represents a vegan, humane, and ecologically friendly response to the animal agricultural tableau depicted here. It also represents an alternative to plant agriculture systems that are premised on the wastes and remains of exploited animals, like the pigs and chickens farmed on this very property, or on chemical fertilizers and synthetic pesticides.

It is common for pig farmers to grow soybeans and corn as feed for the pigs they are raising—while we cannot know for sure in this specific case, this may be what we see in the foreground of *Pig Farm, 2015*. Reimagining this landscape as a veganic one, a diversified plot of vegetables flourishes here instead. Perhaps the veggies are intercropped with a cover crop like alfalfa or clover, which would fix nitrogen in the soil and suppress weeds. An orchard or berry patch, fortified by leaf or hay mulch, surely has a place here as well, perhaps on the far side of the road. And where the structures sit beyond, concealing thousands of living, breathing, feeling, sentient beings who were being prepared for transport at the time this image was made, native grasses and flowers might be planted to sustain free-living animals.

Erasing industrial animal agriculture from this landscape in its transition to a veganic system has broader environmental and public health implications as well. While the pig manure storage system on this farm is not evident, storage and treatment systems can breach and leak fecal bacteria into groundwater. In major storm events, lagoons (open-air pits holding liquid waste) can overflow and contaminate local water supplies. One approach to draining lagoons entails spraying the liquid manure over farm fields; this mist can drift to neighboring properties, coating homes and play equipment. When this waste enters local waterways, the nitrogen and phosphorous can catalyze algal blooms, suffocating aquatic life; pharmaceuticals, antibiotic-resistant bacteria, and parasites are other potential contaminants. Nearby residents endure horrific odors from vast amounts of uncovered or improperly managed waste. Pig manure also releases methane, a potent greenhouse gas. Emissions from chicken barns located elsewhere on this property may include ammonia, hydrogen sulfide, and particulate matter, which can cause or exacerbate asthma and other respiratory ailments in surrounding communities. Chickens and farmworkers are of course exposed to these substances in a concentrated fashion. Prolonged exposure to high concentrations of ammonia can result in eye inflammation and blindness for the birds confined in poultry houses; and chronic exposure to poultry dust links to elevated rates of bronchitis and asthma in chicken-house workers. These represent only a handful of industrial animal agricultural externalities, deleterious to climate, environment, and society, that veganic agriculture avoids.

There are many challenges inherent in transforming a landscape like this, including governmental support for small produce farmers and for good work and fair wages for field laborers; regional cultural norms around animal agriculture; the process of leaving an industry now dominated by contract farming; local marketing opportunities for produce; and the technical knowledge required to adopt a radically different farming approach. Indeed, there is a learning curve for veganic farming, particularly related to soil fertility and product sourcing. While there are various books, websites, and social media pages dedicated to veganic agriculture, the resources available to veganic farmers pale in comparison to those available to chemical-based farmers and the organic farming community. For instance, formal education in veganic approaches does not appear to exist in the US, and extension agents typically do not have sufficient background to advise on veganic approaches. Yet the exploitation and misery concealed on this property as well as the environmental and human health and dignity consequences of these practices beg that we rise to these challenges. A small community of veganic farmers in the United States—and their compatriots around the world—leads this charge.

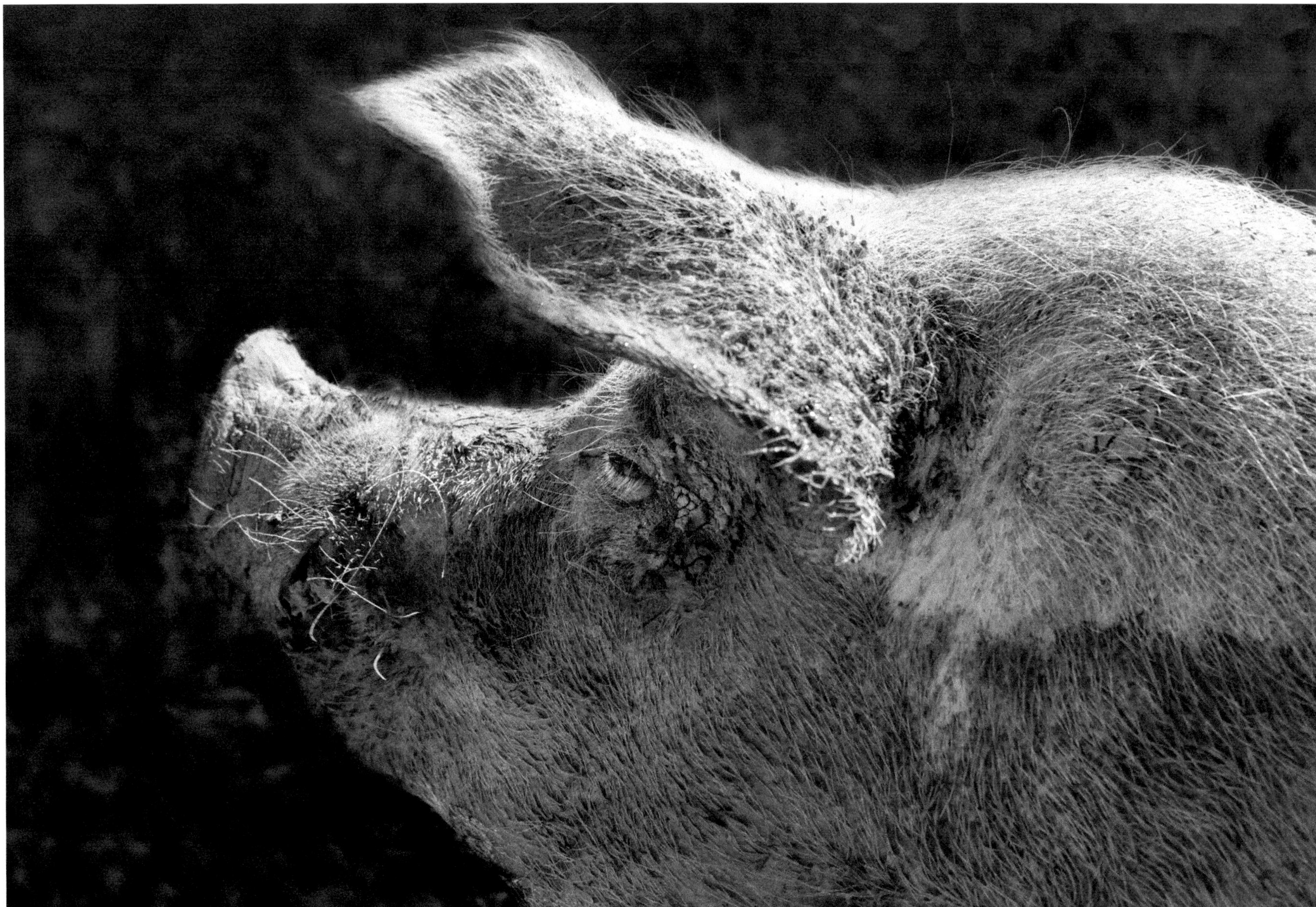

Otto, Escaped from Slaughter, 2023

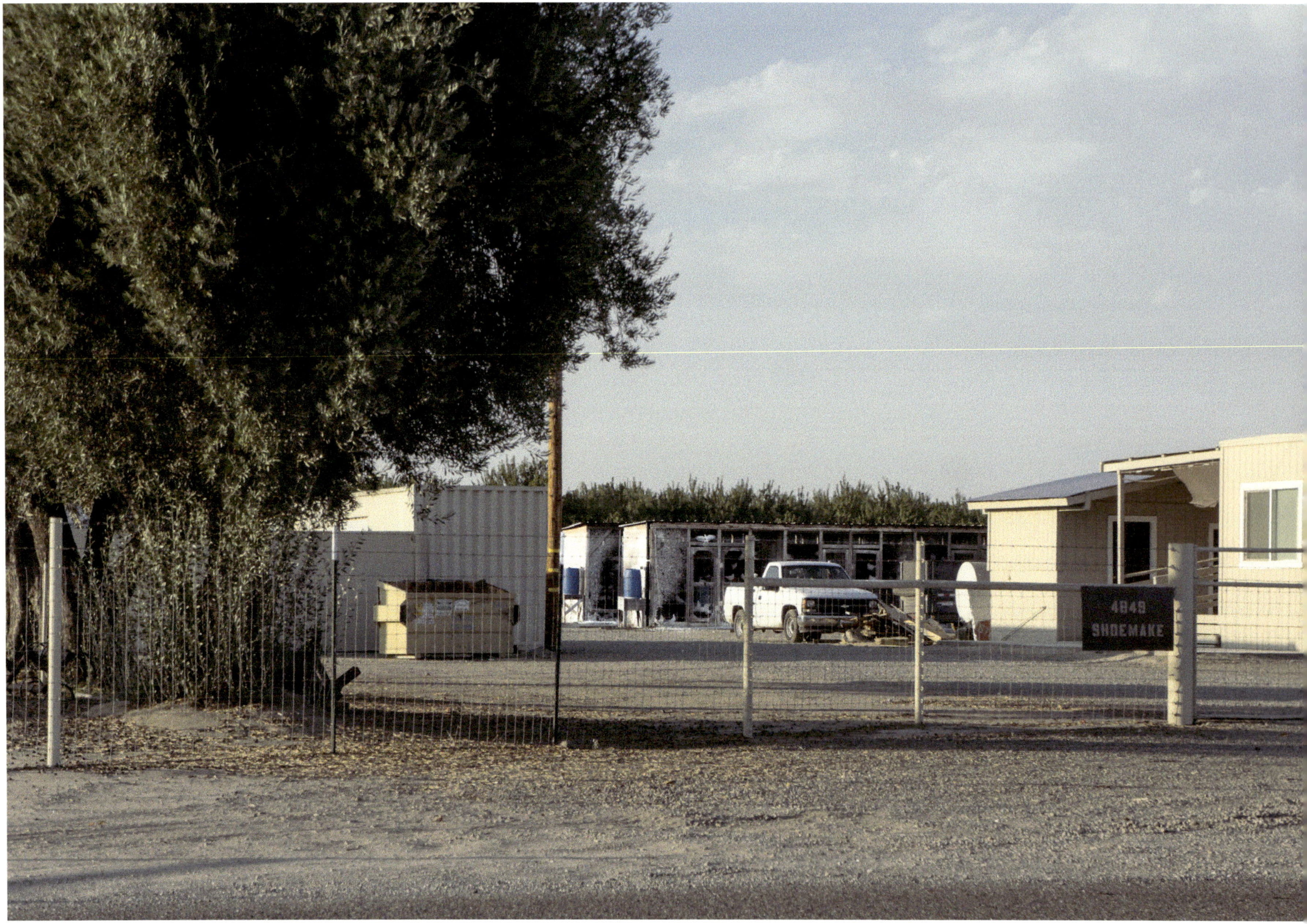
4849
SHOEMAKE

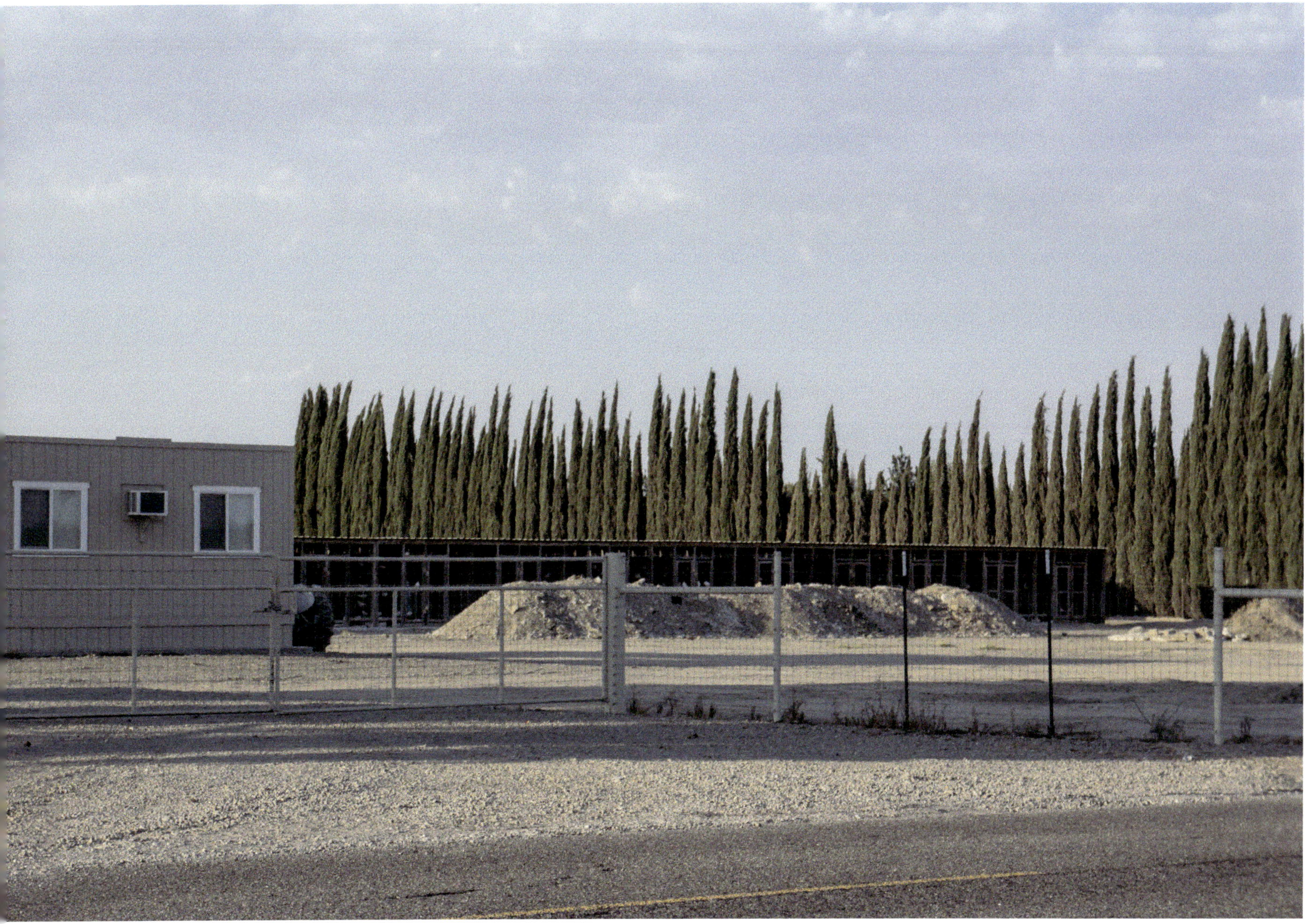

Squab Farm (Pigeons), 2015
6,000

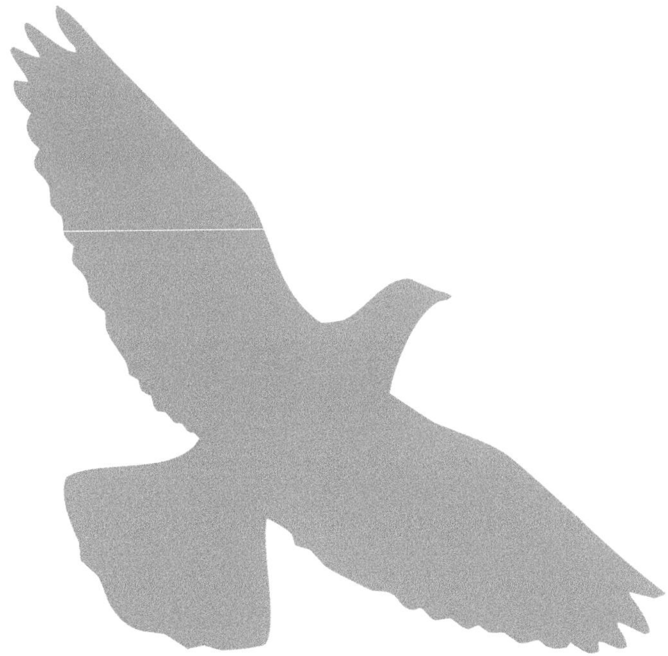

The Persistence of Pigeons

PATTRICE JONES

I can't not see the birds not flying in that sky.

Once so numerous that migrating flocks blocked out the sun, the passenger pigeons who had inhabited North America for 100,000 years were hunted to extinction within 300 years after European pilgrims landed on Plymouth Rock. Passenger pigeons roosted and foraged communally, traveling continuously except when they established temporary "cities" of as many as a million birds each, where they courted, built nests, hatched eggs, and tended the next generation until they were able to fly off with the community.

"Squab" is the inglorious word used by people to refer to the as yet flightless offspring tended so carefully by pigeon parents. Hatched and nursed by both mother and father—who both produce a milk-like substance containing nutrients and antibodies—pigeon nestlings grow quickly, reaching adult size in about a month, at which point they begin to learn to fly. Squab sellers snatch those nestlings just before they fledge.

Flightlessness doomed dodos, ground pigeons who participated in the vibrant pre-human ecology of their island. About three feet high with large curved beaks, dodos foraged for fallen fruits and nuts, including the fruit of the tambalacoque, or "dodo tree." Amazed by the ease with which these gentle birds could be captured, Dutch sailors soon snatched so many that their population plummeted. Dodos died out along with several other endemic species, including tambalacoque trees. Ignorant of almost everything about that destroyed ecosystem, humans added insult to injury by repurposing the word "dodo" as an ableist slur denoting stupidity.

People in the squab business dismissively refer to the doting parents upon whom all of their profits depend as "utility pigeons." Humans have used pigeons as tools in other ways, including as scientific instruments and weapons of war. People also use pigeons as toys, shooting captive birds for sport and relying upon the strong bond between mated pairs (and between parents and offspring) to compel birds bred for speed to race back home through unfamiliar terrain—risking their lives to storms or predation along the way—in the pastime known as pigeon racing.

It all began in the Levant. Neolithic peoples of the eastern Mediterranean built structures for native rock doves to nest, thereby making it easier for themselves to collect droppings to use as fertilizer. The birds who chose to roost in those structures gained access to agricultural spillage in addition to safe housing and were always free to leave. Thus, this might have become one of those mutually beneficial interspecies relationships so often found in nature. Instead, in several places around the region, people stole pigeons from nests to raise in captivity, manipulating their reproduction in order to favor desired features. Some people wanted plump birds to eat. Some wanted birds with feathers of specific colors to be sacrificed in religious rituals. Others valued birds with strong homing instincts and excellent navigational skills to be used for communications.

Successive imperialist empires spread these uses of pigeons around the world. The path from Neolithic towers to this industrial squab farm runs through the Roman Empire, which spread dovecote culture through Europe. After killing off the native passenger pigeons, European settlers to North America imported pigeons for squab farming, the popularity of which has risen and fallen several times in the decades since.

Nobody knows how Mauritius blue pigeons lived. Nobody knows why all of the spotted green pigeons died. People sometimes think they have spotted a Choiseul pigeon, not seen since 1904, but it always turns out to be a crested cuckoo-dove.

None of us know pigeons, not really. The things about them that seem marvelous to us—such as their ability (unique among birds) to make "milk" for their babies or their aptitude for navigation—probably feel entirely unremarkable to them. Objects of endless experiments in countless psychology laboratories, pigeons have proven again and again that they are capable of not only constructing their own communities in a variety of habitats but also solving artificial problems posed for them by apes in lab coats. But such tests tell us nothing about what pigeons themselves want, think, or wonder.

We can guess, from their behavior, what pigeons care about and what they enjoy. Pigeons value relationships. They mate for life. They protect, indulge, and instruct their young. They socialize constantly with their flock mates (coo coo coo) and often organize exuberant collective aerial acrobatics. Judging from the evident joy that they take in those recreational flights, pigeons also value physical freedom and communal activity.

What are their other projects? We cannot know.

When ornithologist Johann Reinhold Forster spotted what turned out to be the last of the ground doves of Tanna, called "mahk" by the native people of that island, he shot her.

People are the primary predators of pigeons. They must know us better than we ever will know them. Since their views of people are undistorted by the rose-tinted spectacles of speciesism, they may know us better than we know ourselves.

What do pigeons know about people? Remember: In Pennsylvania, politicians raise money by staging events at which newly-fledged captive-bred pigeons are released for the first time—in order to be shot from the sky. Psychologists compel pigeons to solve problems by withholding food, sometimes deliberately mutilating their brains before playing such games. Pigeon fanciers proclaim their love for favored birds and then sell their offspring to the highest bidder.

What might we learn about ourselves or others by imagining pigeon perspectives? What do the preoccupations of people look like from those bird's-eye views? What might the 6,000 birds locked up in the buildings in that photograph tell us that we urgently need to know?

While Rodrigues blue pigeons, Bonin wood pigeons, and Réunion pink pigeons are as extinct as passenger pigeons, dodos, spotted green pigeons, Choiseul pigeons, and Tanna ground doves, more than 300 species in the pigeon and dove family have survived the onslaught of the human, often exercising great ingenuity in so doing. They include wood pigeons, ground pigeons, mountain pigeons, and fruit doves as well as the rock doves who are the living ancestors of all of the human-engineered varieties of captive pigeons and doves. The colors of their feathers span the spectrum of the rainbow.

Pigeons persist. If a predator snatches their nestlings, they try again. And again. If their habitats are razed, they find ways to nest in new places. Those grey-and-black pigeons you see in cities? The ones who seem so pedestrian and are often perceived as pests? They're busy reclaiming the means of production, by which I mean their own bodies: these birds are the descendants of pigeons released from dovecotes by the thousands when squab farming became less profitable, as well as escapees from this and other exploitive industries.

In Christchurch, New Zealand (where the wood pigeons called kererū are lauded for their habit of getting drunk on fermented fallen fruit, but the feral descendants of pigeons imported for squab production may be shot on sight), I have seen pigeons flying in and out of the homes they have made for themselves in the wreckage of buildings brought down by an earthquake.

In Rome, Italy (from whence the Roman Empire launched pigeons around the world), I have seen the self-possession of pigeons regarded by visitors as whimsical tourist attractions. Perhaps these Indigenous Mediterraneans, who are well-prepared to adapt themselves to whatever climate change may bring, know that they will be the inheritors of the future ruins of all of those fountains and cathedrals.

In Springfield, Vermont (where passenger pigeons used to soar), I see the rewilded offspring of pigeons rescued from hunting, racing, and squab production swooping around the grounds of VINE Sanctuary every morning. Their descendants will survive me. I am heartened by their future feathers.

I can't not see the birds not flying in that sky. But I know that nature abhors a vacuum, including the amorality and absence of empathy in the hearts of the owners and operators of that site of concentrated cruelty. Someday, one way or another, it will be gone. But pigeons will persist.

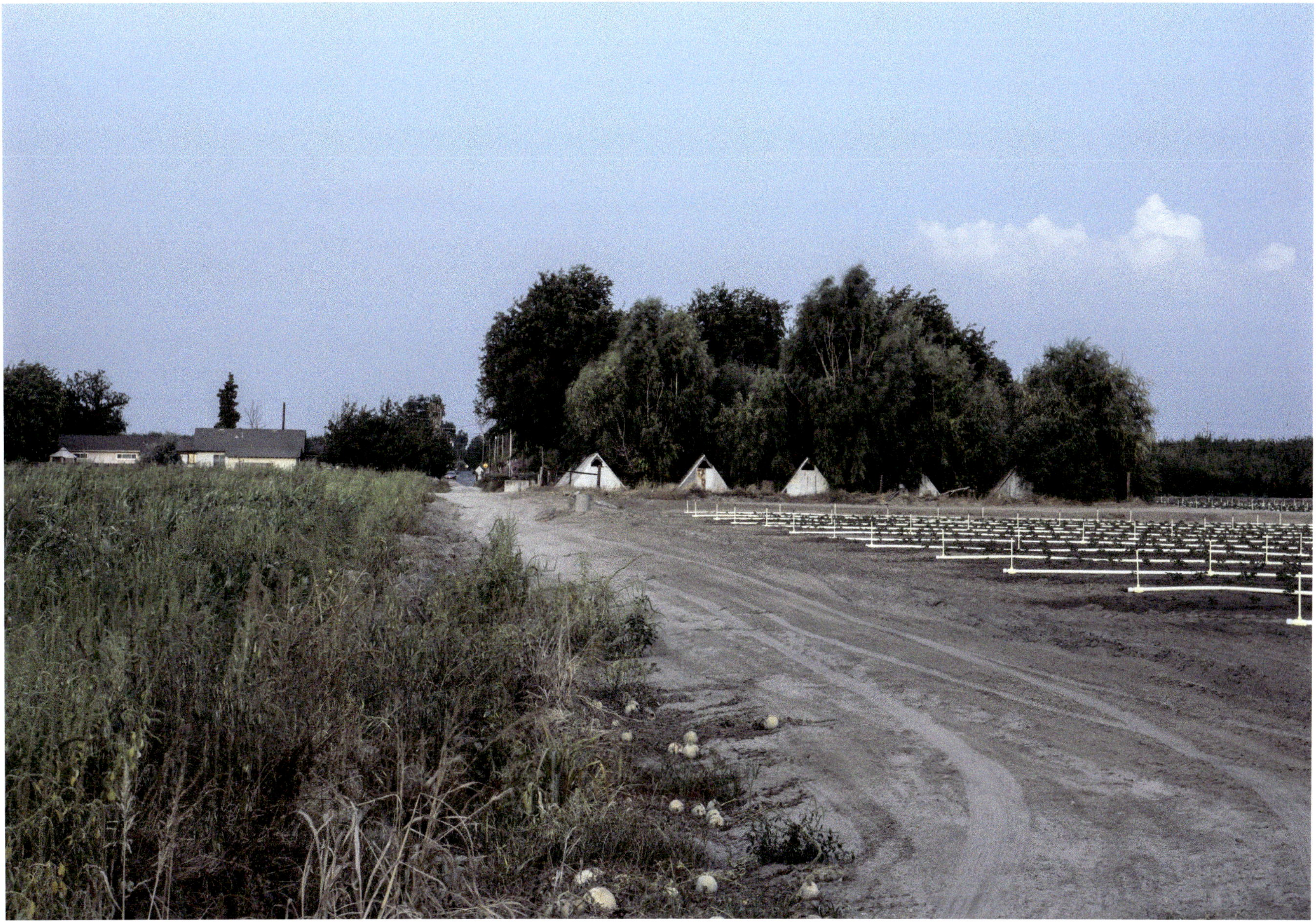

Rabbit Farm, 2015
3,000

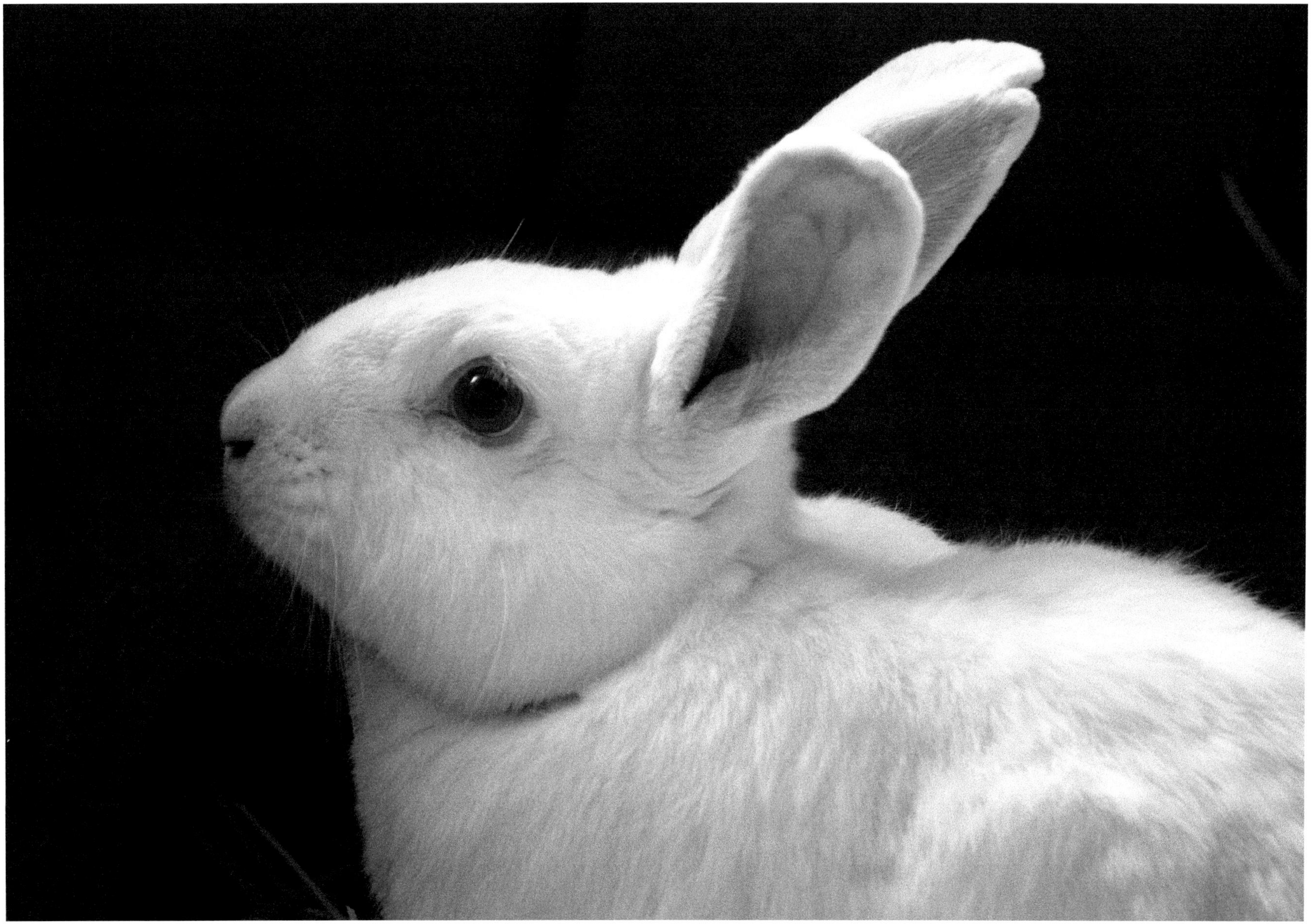

Marcus—Rabbit Farm Rescue, 2014

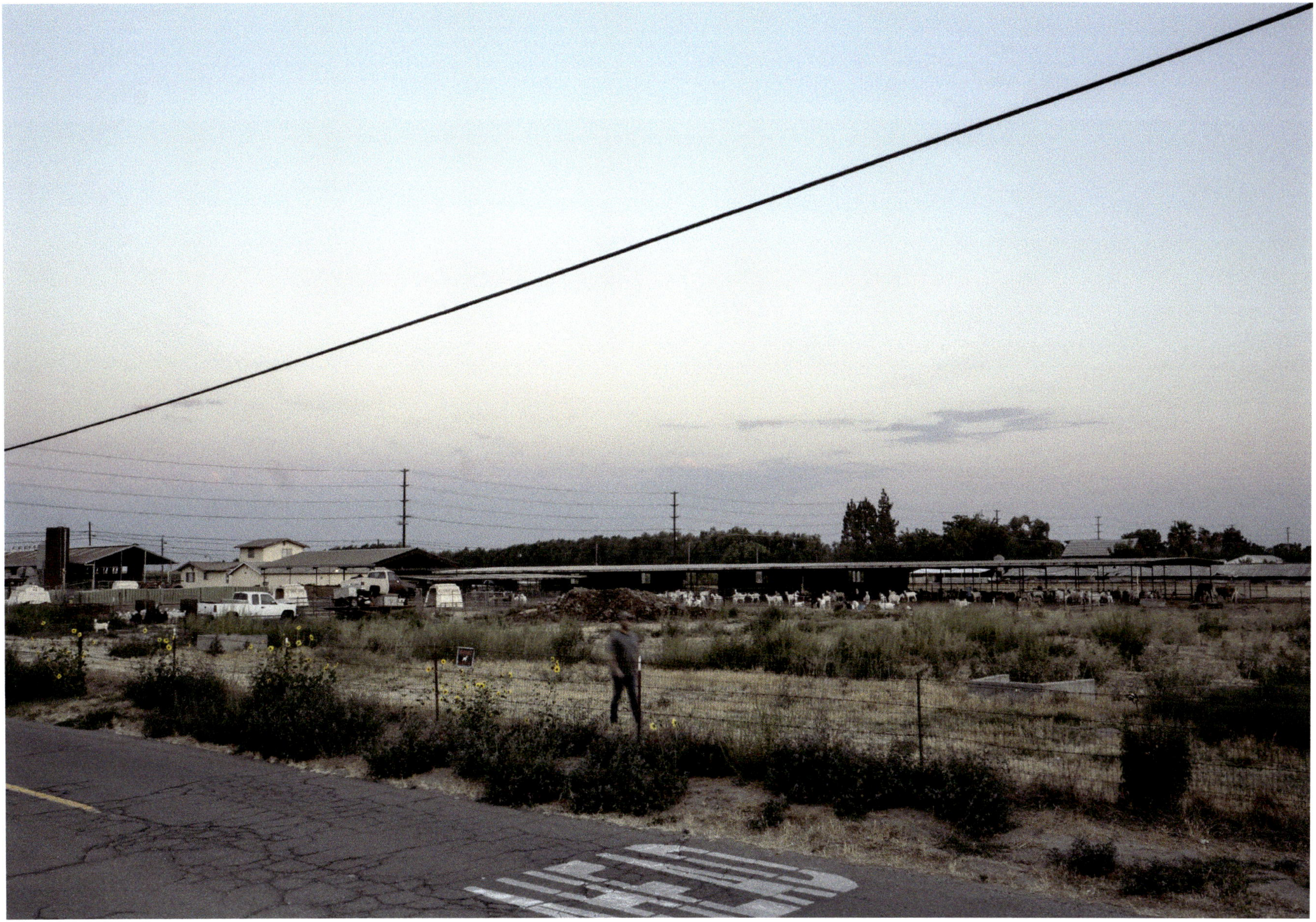

Dairy Farm (Goats and Cows), 2015

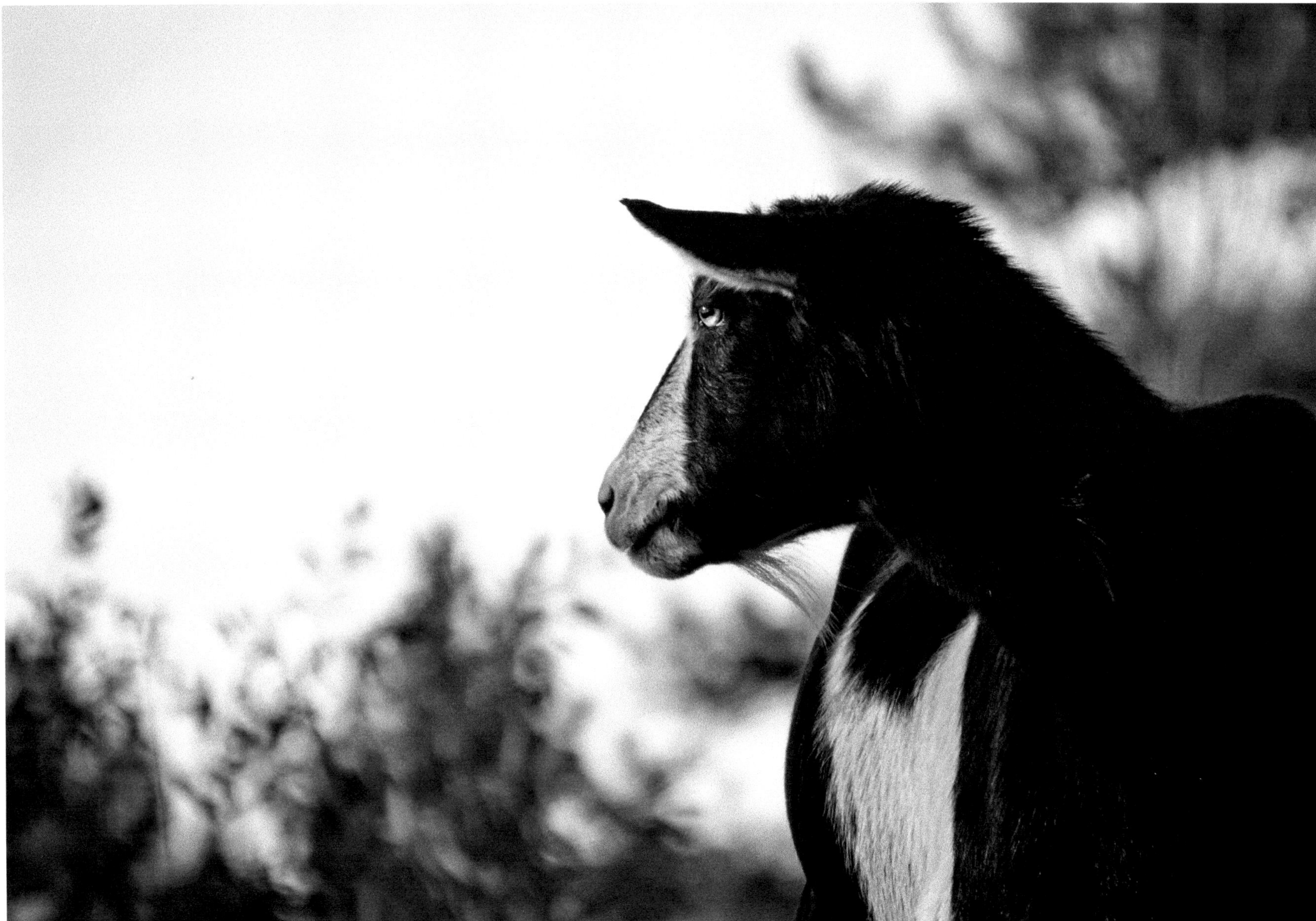

Stella—Dairy Farm Rescue, 2016

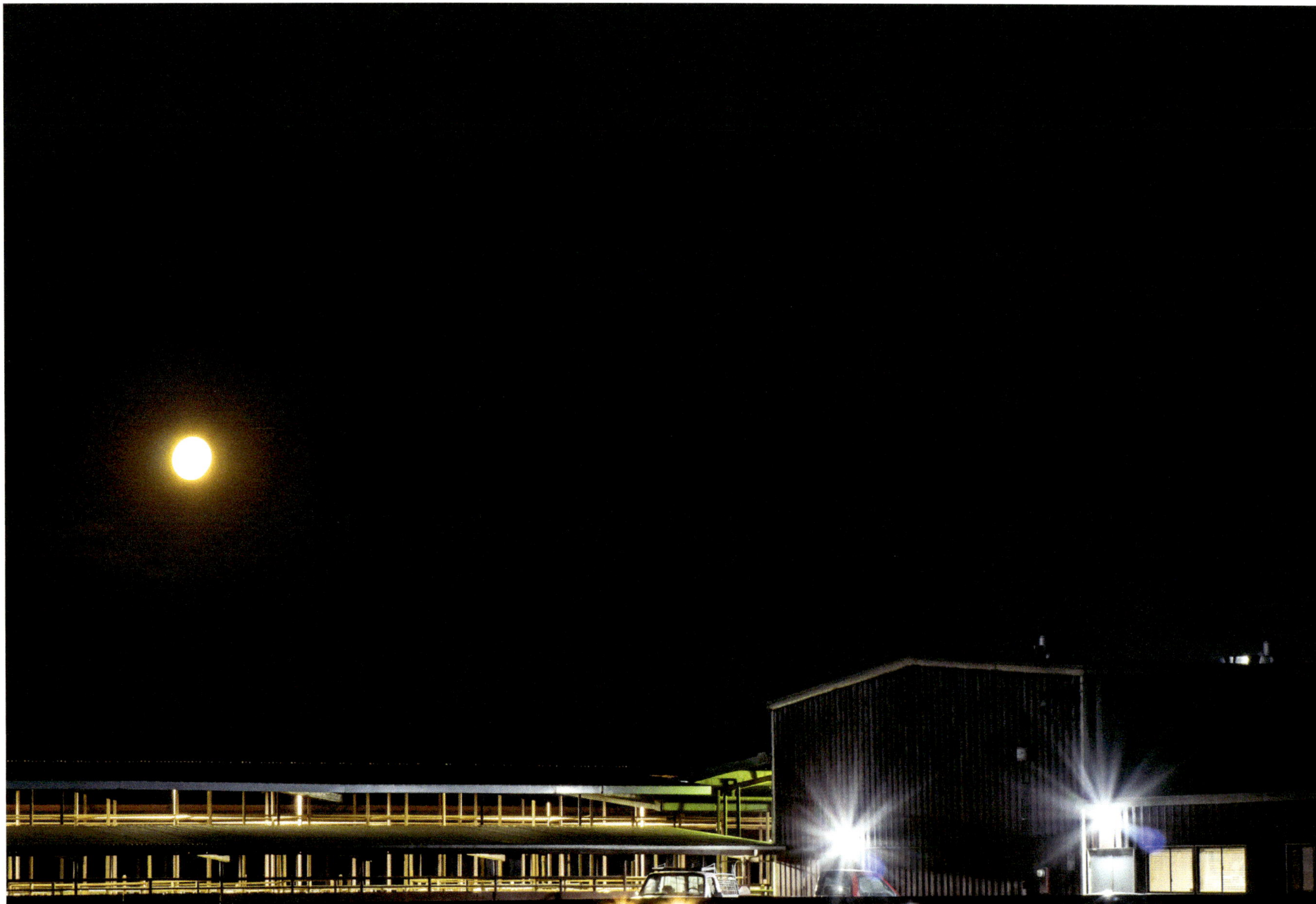

Night of the Full Moon Livestock Auction (Cattle), 2015
150,000/year

Of Cash Cows and Golden Eggs FACTS

FACT: Between 1995 and 2020, livestock operators received an average of almost $2 billion a year from the US Department of Agriculture. This does not include livestock commodity payments, dairy subsidies, or research grants.[1]

FACT: Between 2018 and 2022, the USDA spent less than .5 percent of that a year (an average of $10 million a year) to support plant-based and other alternative proteins.[2]

FACT: Between 1995 and 2020, the USDA paid an average of over $6 billion a year subsidize farmers who grow animal feed like corn and soybeans.[3]

FACT: In 2020, the USDA paid over $52 billion in agricultural subsidies.[4] $11 billion (over 20 percent) of federal agricultural subsidies were paid to livestock operators.[5]

FACT: In 2020, only 2.5 percent to 5 percent of US federal farm subsidies went toward fruits and vegetables.[6]

FACT: In 2018, just five poultry companies—Tyson Foods, Pilgrim's Pride, Sanderson Farms, Perdue Foods, and Pilgrim's Pride—controlled over 60 percent of production in the US.[7]

FACT: In 2021, Tyson Foods paid $221.5 million to settle accusations of colluding with competitors to raise chicken prices.[8]

FACT: In 2021, Pilgrim's Price pleaded guilty to price-fixing charges and paid $107 million in fines.[9]

FACT: In 2022, the president/CEO and the chairman of the board of Tyson Foods received over $12 million each in compensation.[10]

FACT: As of June 2023, the average pay for a poultry worker in the US was $15.91 per hour.[11]

FACT: In 2020, the four largest dairy co-ops controlled over 50 percent of the US market.[12]

FACT: Because of the 1922 Capper–Volstead Act, dairy co-ops are exempt from many antitrust laws.[13]

FACT: In 2013, Dairy Farmers of America (DFA) paid $158.6 million to settle a lawsuit alleging milk price-fixing in the southeast United States.[14]

FACT: In 2016, DFA paid a $50 million settlement in a lawsuit alleging an illegal conspiracy to restrain competition, fix and suppress prices, and control the sale of raw milk in the northeast United States.[15]

FACT: In 2020, DFA settled another lawsuit, for an undisclosed sum, with a separate group of northeastern US dairy farmers.[16]

FACT: From 2009 to 2017, during the Obama administration, and again starting 2021, during the Biden administration, Thomas J. Vilsack has been the US Secretary of Agriculture. From 2017 to 2023, he was also president and CEO of the US Dairy Export Council.[17]

FACT: In 2023, the US dairy industry paid $7,237,503 for government lobbying.[18]

FACT: A survey conducted in 2014 and 2015 of eighty-eight immigrant dairy farmworkers on fifty-three different dairy farms in New York state reported that these workers earned about $9 per hour and worked twelve hours a day with few breaks and no overtime pay. Ninety-seven percent lived in on-farm housing, the majority of which were infested with insects.[19]

FACT: In 2014, immigrant labor accounted for 51 percent of dairy labor and produced 79 percent of the US milk supply. Dairy farm workers were compensated an average of $11.54 per hour including benefits.[20]

FACT: The US Bureau of Land Management administers 18,000 permits and leases to ranchers to graze their livestock on 155 million acres of public lands. The permits and leases generally cover a ten-year period and are renewable.[21]

FACT: In 2021, the four largest meat-packing companies—JBS USA, Tyson Foods, Cargill Inc, and National Beef Packing Company—slaughtered about 85 percent of US beef.[22]

FACT: In 2021, the four largest pork companies—Smithfield Foods, JBS USA, Tyson Foods, and Sanderson—controlled 70 percent of the US market.[23]

FACT: In 2011, the CEO of Smithfield Foods received almost $19 million in compensation.[24]

FACT: In 2023, the median pay for slaughterhouse workers was $13.76 per hour.[25]

FACT: In 2022, a US federal lawsuit accused meat industry giants, including JBS USA, Cargill Inc, and Tyson Foods, of conspiring to keep employee wages low.[26]

FACT: In 2021, the US dairy industry paid $6.8 million for government lobbying. In 2022, $7.3 million.[27]

FACT: In 2021, the US meat industry paid $4.5 million for government lobbying. In 2022, $4.12 million.[28]

FACT: In 2020, 37.5 percent of workers in animal slaughtering and processing were foreign-born. An unknown but substantial number of workers in this industry are undocumented. In one of the most dangerous industries, they cannot advocate for their own safety, health, fair compensation, or benefits for fear of losing their job or deportation.[29]

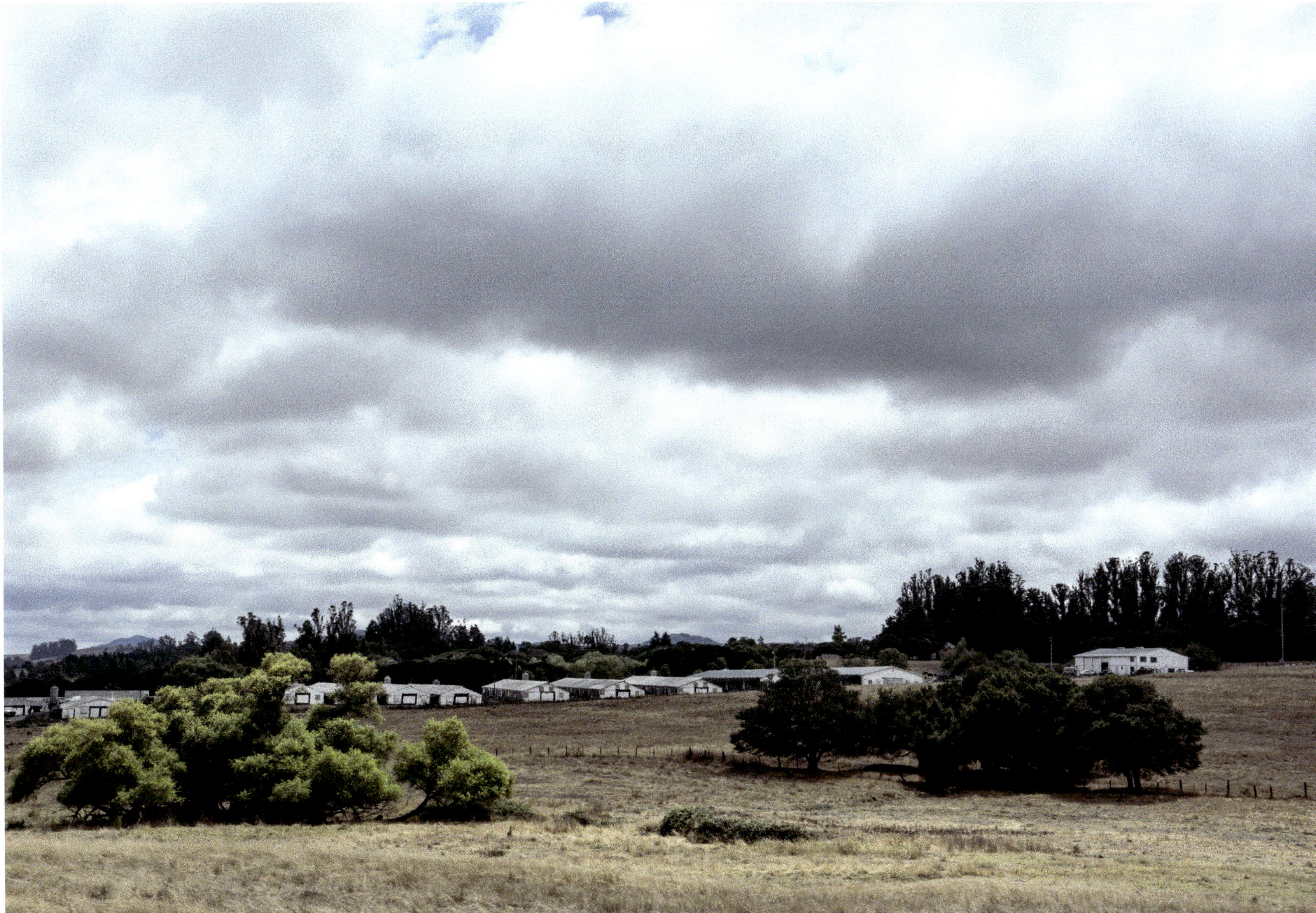

Organic Chicken Farm, 2015
125,000

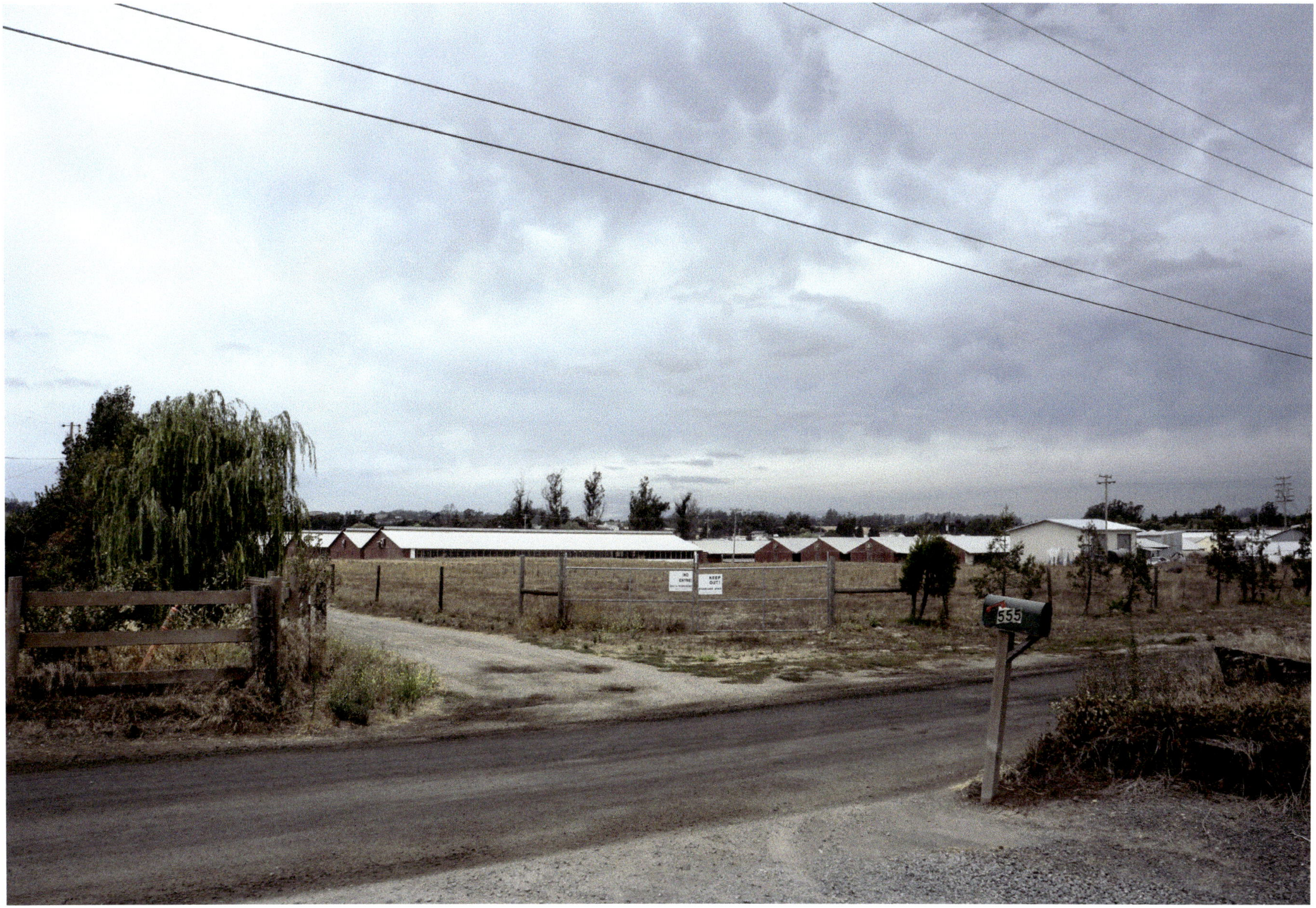

Organic, Cage-Free, Humane Certified Egg Farm (Chickens), 2015
250,000

Eating Animals and Human Health

- Elevated levels of hydrogen sulfide, ammonia, particulate matter, and other toxins in Concentrated Animal Feed Operations (CAFOs) have been linked to serious respiratory illnesses, including chronic bronchitis, asthma, and sinusitis, in workers and nearby communities.[1]
- CAFO workers have an increased risk of musculoskeletal disorders and loss of hearing from heightened noise levels. In 2003, the American Public Health Association issued a recommendation urging "federal, state and local governments and public health agencies to impose a moratorium on new CAFOs until additional scientific data on the attendant risks to public health have been collected and uncertainties resolved." There has been little government response.[2]
- As of 2015, 62 percent of antibiotics are used on farmed animals in the US.[3] High use of antibiotics results in antibiotic-resistant superbugs such as Campylobacter, MRSA, and Salmonella. Over 2.8 million antibiotic-resistant infections occur in the US each year, and more than 35,000 people die as a result.[4]
- Marine animals are exposed to PCBs (polychlorinated biphenyls) that pollute our waterways. Farmed salmon are fed fishmeal made from ground small fish, which causes additional PCB accumulation. Farmed salmon are likely the most PCB-contaminated protein source in the US food supply—up to forty times more than other foods. PCBs cause cancer and were banned in the US in 1976.[5]
- Increased rates of diabetes,[6] heart disease, stroke,[7] and cancer[8] have been associated with eating red and processed meat.
- Higher rates of obesity are associated with eating meat,[9] including poultry.[10]
- Eating meat, including poultry, is associated with an increased risk of developing diabetes.[11]
- Eating eggs increases the risk of heart disease and cancer mortality.[12]
- An estimated 68 percent of the world's population is lactose intolerant.[13]

- Dairy products have been associated with autoimmune diseases, including diabetes, celiac disease, inflammatory bowel disease, rheumatoid arthritis, multiple sclerosis, and schizophrenia.[14]
- Countries with the highest rates of dairy consumption have the highest rates of osteoporosis-related hip fractures.[15]
- Consuming dairy products has been associated with increased risk of prostate cancer.[16]
- Consuming dairy milk has been associated with increased risk of ovarian and breast cancer.[17]
- Dairy products are associated with acne in individuals seven to thirty years old.[18]
- In 2013, 48.7 percent of corn grown in the US[19] and over 70 percent of soybeans were fed to farmed animals.[20] These two crops account for half of US farmland.[21] Herbicide use by 1980 reached 90–99 percent of US corn and soybeans.[22] These crops also depend heavily on pesticides. Exposure to pesticides and herbicides has also been linked to a variety of health problems in humans including reduced fertility, low birth weight, neurological disorders, numerous types of cancer including brain, pancreatic, non-Hodgkin lymphoma, myeloma, colon, testicular, and soft tissue sarcoma.[23] International affairs think tank Chatham House lists change to predominately plant-based diets as the first of three key levers in support of biodiversity and improved public health.[24]
- A plant-based diet protects against breast cancer, prostate cancer, colorectal cancer, gastric cancer, pancreatic cancer, and lung cancer.[25]
- A plant-based high-carbohydrate/high-fiber diet is more effective against type 2 diabetes than a high-monounsaturated fat/low carbohydrate diet.[26]
- A plant-based diet is recommended in the prevention and treatment of obesity.[27]
- A plant-based diet is associated with a lower risk of heart disease and with longer life.[28]

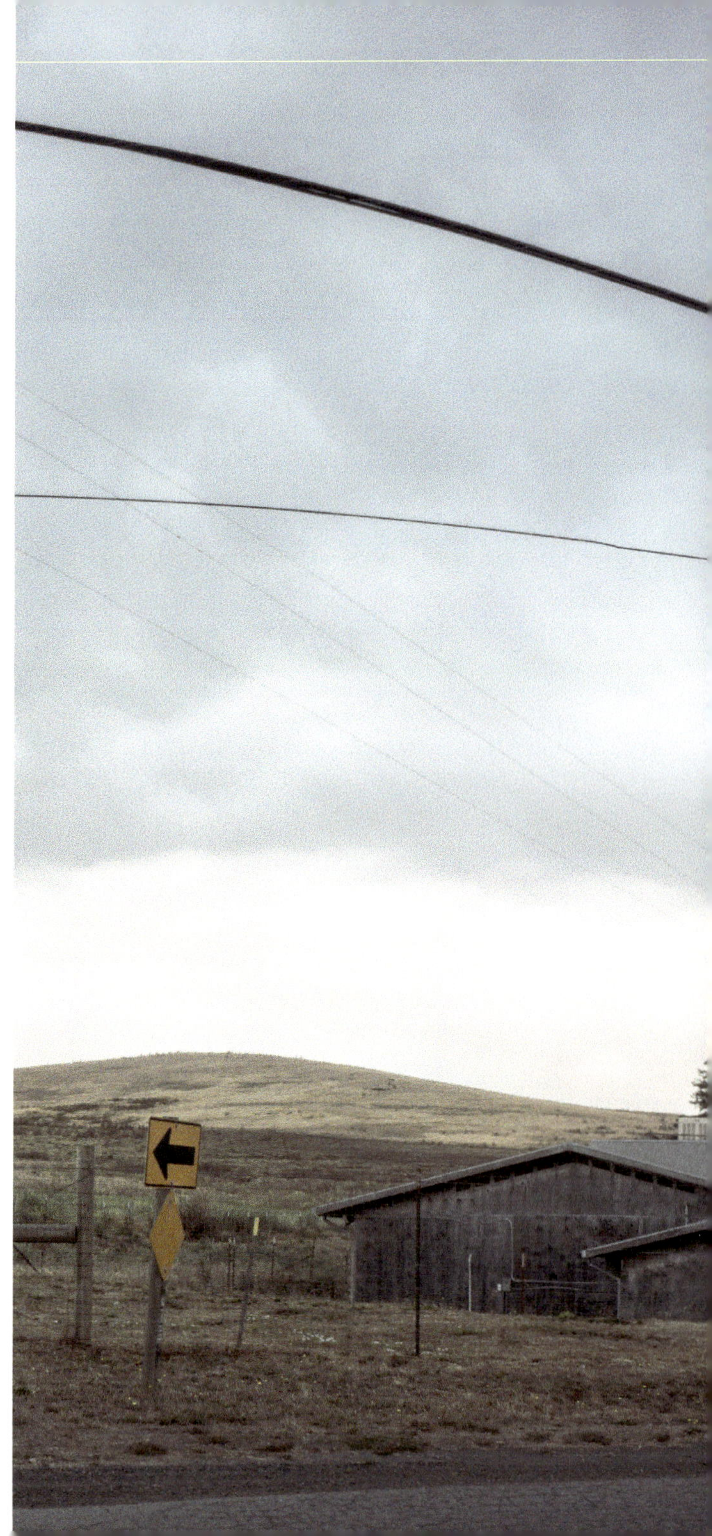

Duck Farm, 2015
300,000

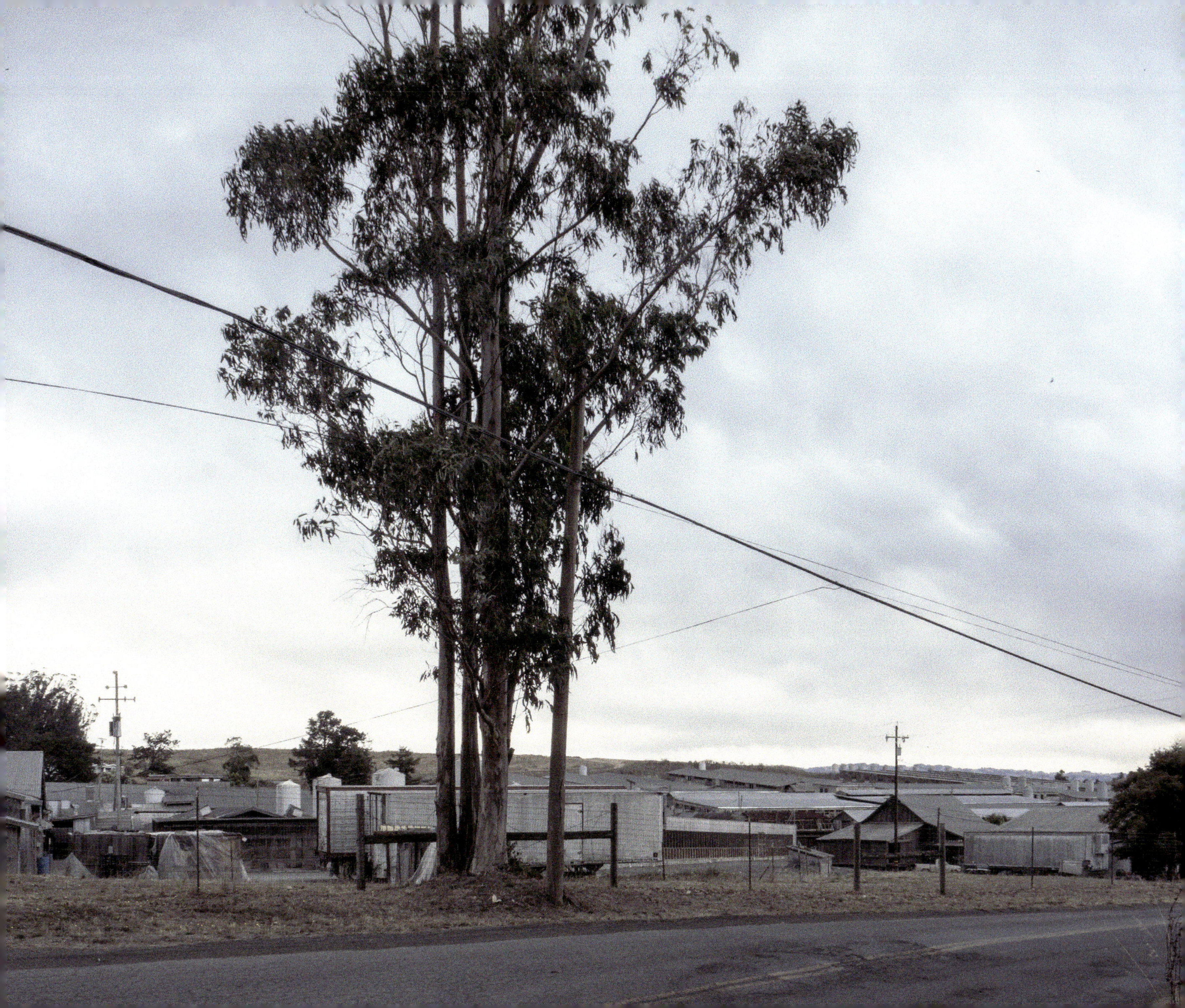

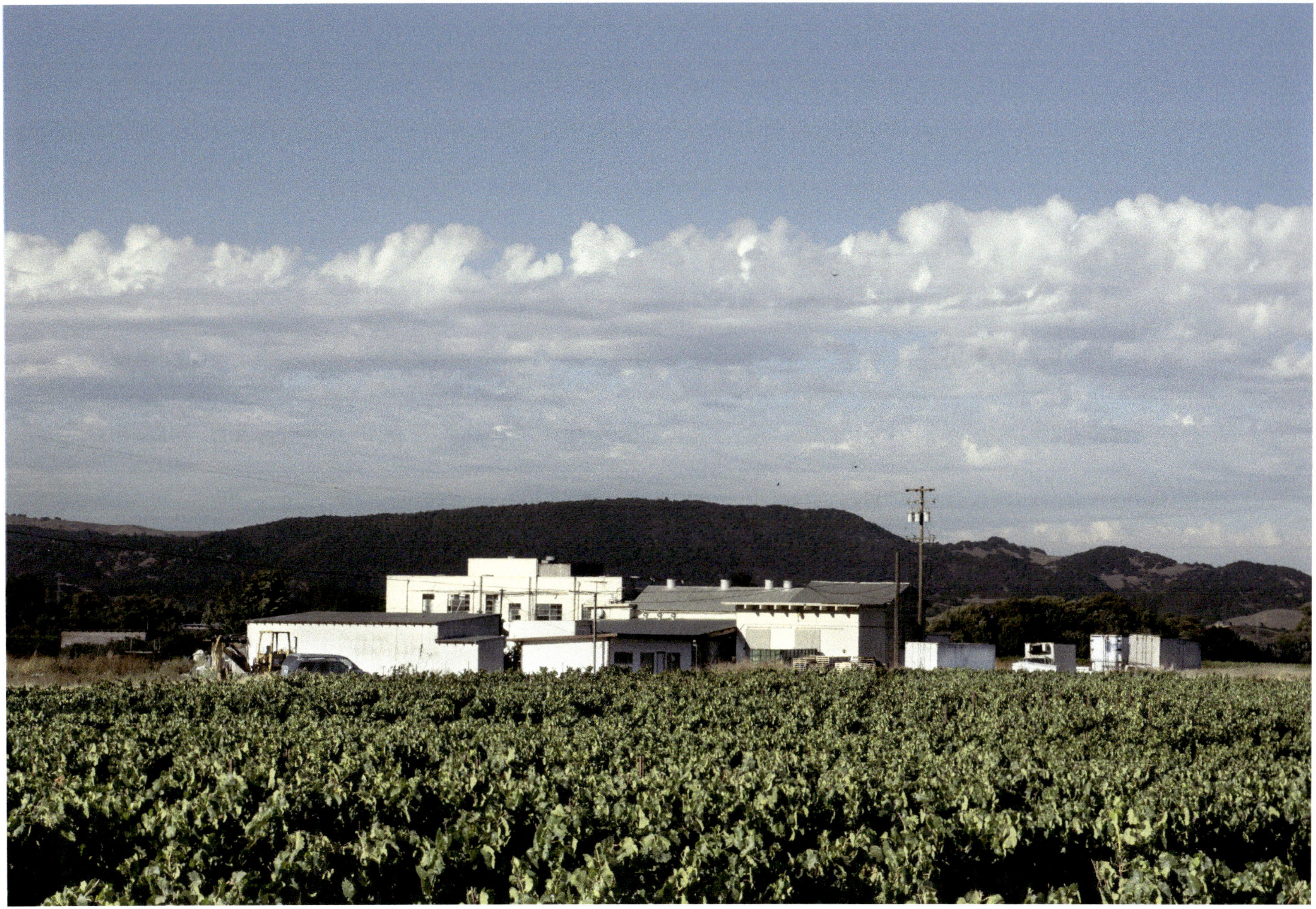

Slaughterhouse (Turkeys), 2015
85,000/year

E.T. Phone Home

I have had contact with extraterrestrials. No, I'm not a kook.

They are complex creatures, sometimes repulsive to *Homo sapiens* but sublime even to the limited perceptions of the human eye. Their senses are often more developed and acute than those of humans. They communicate with sophisticated vibrations humans are incapable of understanding. They are intelligent beings.

When I was younger, I contributed to their mass slaughter. I deeply regret this. E.T.s feel pain, and my actions caused immeasurable suffering.[1]

The torment continues. Humans imprison them, breed them, farm them, and slaughter them with impunity—with glee even. Humans kill at least a trillion E.T.s every year.[2] Some humans boil them alive. Some humans eat them alive. Humans, usually men, bait them with lures on hooks that pierce their flesh, for sport. The sportsmen bond with their children over this pastime. Women who pursue this activity are touted as feminist icons.[3]

The SETI (Search for Extraterrestrial Intelligence) Institute is a non-profit organization founded in 1984. Great minds have devoted themselves to, and millions of dollars have been spent on this kind of research. In 2020, the SETI Institute helped establish the Earthling Project to create a composition of 10,000 voices representing the sounds of humanity that will be sent to outer space.[4]

Meanwhile, the sounds of humanity are ruining extraterrestrial habitats. Boats, ships, motors, SONAR, and offshore blasting disrupt E.T.s' ability to communicate, to find food or mates, and to avoid predators, causing trauma or death.[5]

The industries of humanity are destroying extraterrestrial habitats. At least fourteen million tons of plastic every year end up in their home. It poisons them. It chokes them. It causes starvation.[6] Agricultural runoff containing pesticides, fertilizers, and farmed animal feces produces harmful algal blooms and marine dead zones.[7] Farmed E.T.s infect wild E.T.s with lethal diseases.[8]

Humans long ago encountered extraterrestrial intelligence. The consequences have been tragic.

> *Dear Extraterrestrial Beings,*
> *Beware of* Homo sapiens *from planet Earth!*
> *Danger, danger!*

*"Extraterrestrial" originates from the Latin *extra terrestris*, which means "outside of soil, land, ground, or earth."

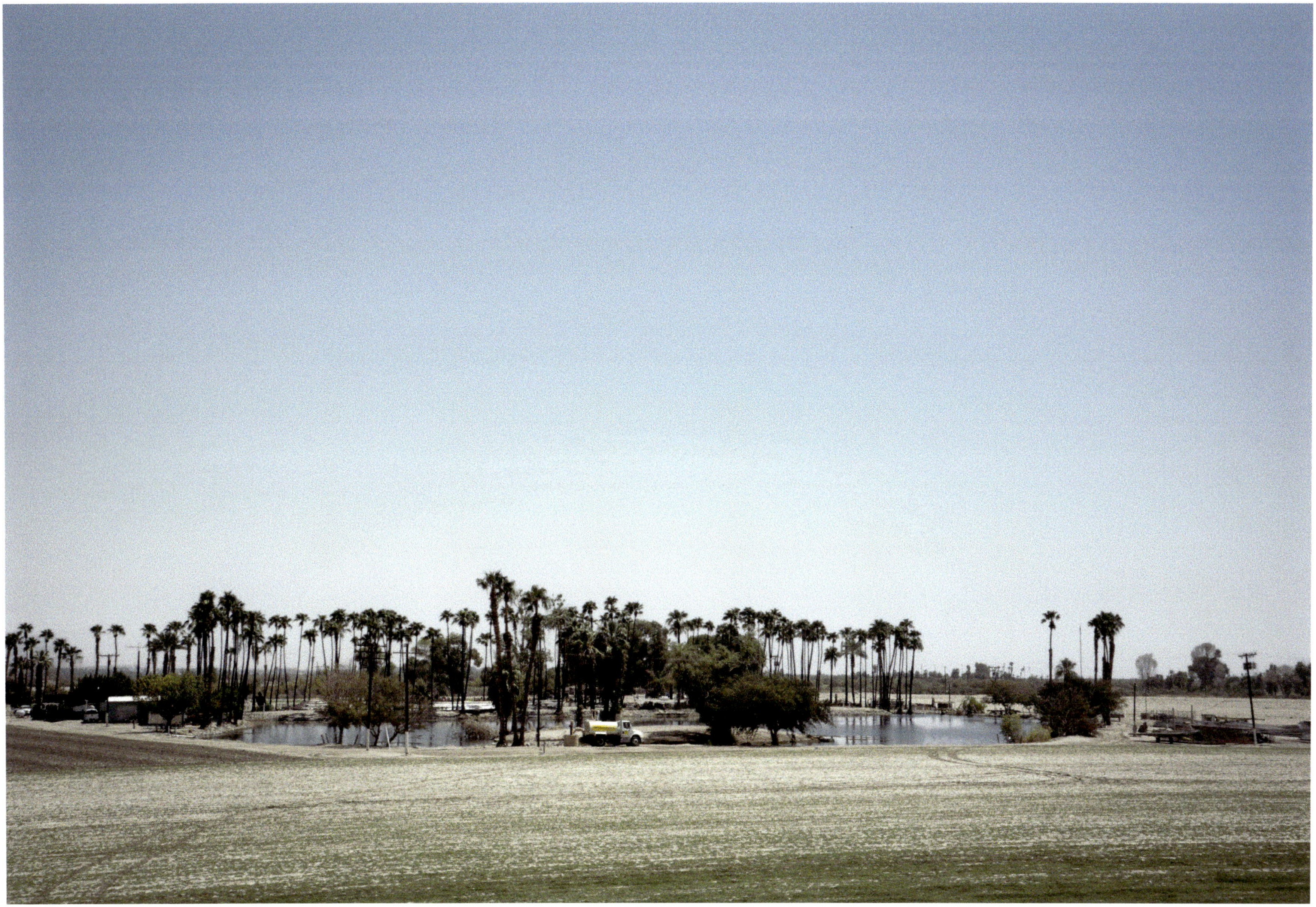

Fish Farm, 2015
3,000,000

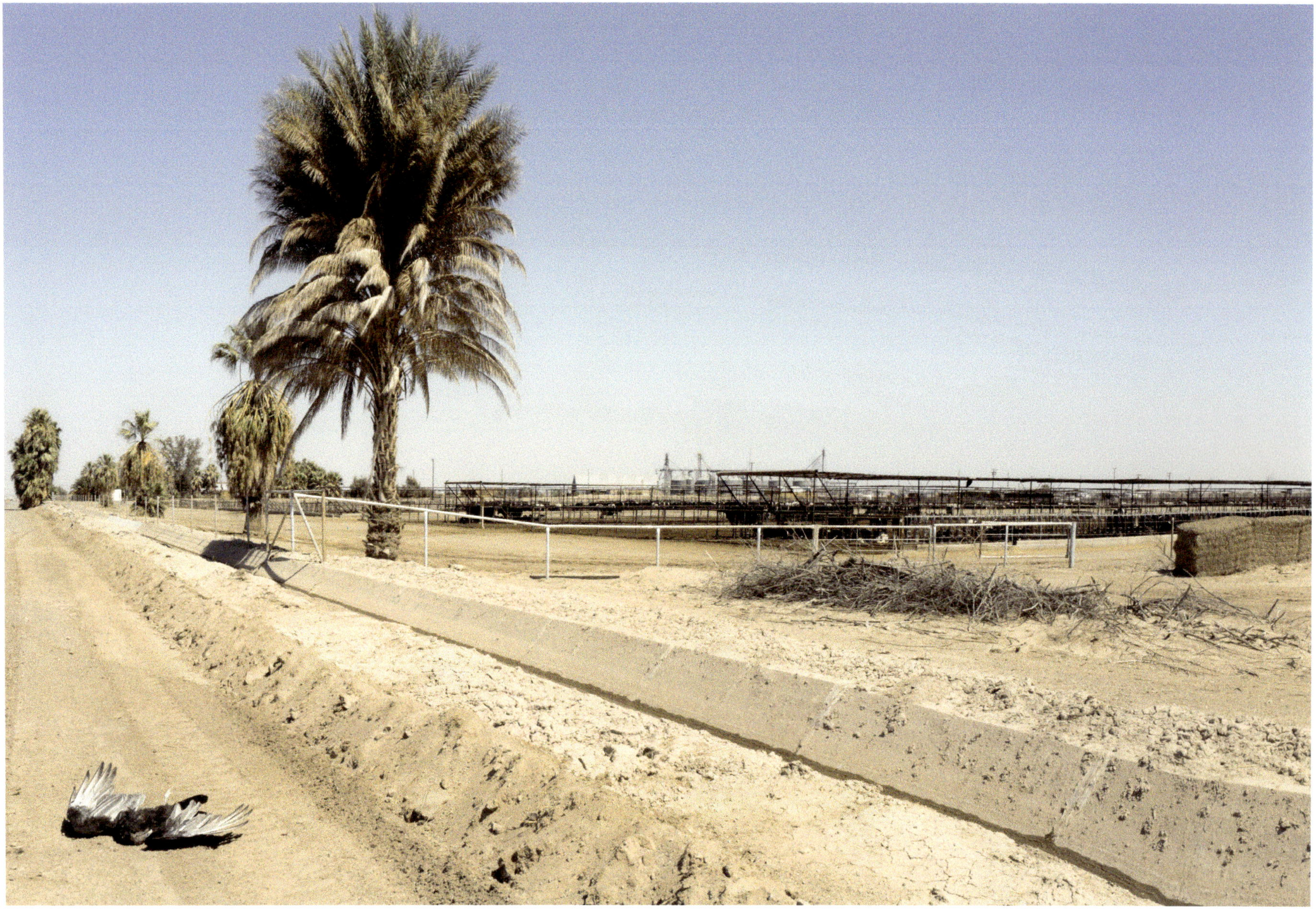

Feedlot (Cows) 116 Degrees Fahrenheit, 2015
15,000

Decolonize/Re-Indigenize

By the time I turned eleven years old, my family had moved ten times from El Paso, TX, to Albuquerque, NM, back to El Paso to Bedford, MA, to Brescia, Italy, to Framingham, MA, to Queens, NY, to Omaha, NE, to Somerville, NJ, to Audubon, PA. Why did we move so much? One factor is that my immigrant parents were chasing the American Dream. Evidence suggests frequent moving is not unusual for immigrant families.[1] By age eighteen, I was on my own, and had and moved sixteen more times for a total of twenty-six different homes.

Who knows for certain how all those good-byes and always being the new girl affected me. My grip on relationships may be more tenacious than average, but perhaps that's just my nature. I regard friends who grew up in the same house until they became adults with some envy and a skosh of superiority; surely, they couldn't have had as many adventures. My first language was Spanish, but by the third grade I spoke English almost without an accent. To try to fit in and not get picked on as different or other, I grew ultra aware of regional accents and of the verbal and unspoken tics and expressions of people around me. I became subconsciously adept at what is now called code switching. So much displacement nurtured in me an obsession with origins and roots, especially the indigenous roots of the places I've lived—native plants, native wildlife, Indigenous peoples.

If I could cast roots down through time on the land where I now stand, before colonization, I would find myself among people referred to as Shawnee. And if I could send roots to the time before colonization in all the homes where I've lived, the people I would find myself among would include Ohlone, Susquehannock, Piankeshaw, Canarsee, Lenni Lenape, Omaha, Nipmuc, Massachusett, Jumano, Pueblo, Navajo, and Mescalero. My deepest American roots, the ancestral roots in my blood and DNA passed to me from my Mexican mother, are at least in part those of the Rarámuri in northern Mexico.

Like many people of Indigenous heritage, I can't be sure of my lineage. The exaltation of the European settler colonizers at the expense of Indigenous peoples blurred or erased the history of Indigenous communities. The names First Nations identified themselves with have often been misunderstood and mangled by the Europeans and etched into the colonialist version of history that prevails in American schools. Through violence, genocide, deracination, enslavement, and the introduction of deadly infectious diseases, the European colonizers were responsible for the death of 90 percent of the Indigenous

American population.[2] The colonists decimated Indigenous languages, histories, beliefs, customs, and agriculture.

Diverse First Nations thrived in the Americas for thousands of years before the Europeans arrived, in no small part due to advanced systems of agriculture. Unlike most modern farming, pre-Columbian agriculture integrated itself into local ecosystems and is now understood by contemporary ecologists to be regenerative and agroecological.[3]

Indigenous peoples in North and Central America did not farm animals. They produced tasty, nutritious food for themselves without breeding or caging other animals. Cows, pigs, chickens, goats, and sheep did not exist in the Americas; farmed animals were brought by the Europeans,[4] along with alcoholism,[5] epidemics,[6] and firearms.

The colonial practice of breeding animals to feed humans is inefficient, resource intensive, and environmentally catastrophic (see "Environmental Impacts"). For instance, farming animals requires much more water than raising plants for food,[7] which contributes to the drought that has gripped the American West,[8] which contributes to increased wildfires,[9] which contribute to climate change, which contributes to drought and wildfires and so on and so on in a vicious cycle.[10] Indigenous Americans manage wildfires with controlled burns and fire breaks.[11] Communities living in arid climates have dug and maintained complex systems of canals for irrigation. In humid climates they've planted on mounds to drain off excess water.[12]

In contrast to the diminishing biodiversity of current American agriculture, much of which is fed to farmed animals (see "Environmental Impacts"), Indigenous Americans have cultivated a multitude of plants, including tomatoes, avocados, cacao, chia, peanuts, papaya, wild rice, potatoes, peppers, amaranth, and quinoa.[13] Over the course of at least ten centuries, long before Columbus and Cortés arrived, the cultivation of maize, also known as corn, spread throughout the Americas. First Nations who lived in agricultural areas considered maize sacred.[14] Indigenous Americans practice intercropping, planting together the Three Sisters: squash, maize, and beans. These three crops support, strengthen, and protect each other. Squash spreads along the ground, preventing weeds, retaining moisture, and deterring pests with prickly leaves and stems. Maize grows into a climbing structure for the bean plants, which in turn secure the maize in high winds. Beans fix nitrogen, which feeds the soil. Together they form a resilient, biodiverse ecosystem more efficient at using light, water, and nutrients and less vulnerable to pests than monocultures.[15] Nutritionally, the Three Sisters balance each other, providing a diet rich in complete protein and vitamins.[16]

Indigenous agricultural practices have proven to mitigate climate change. For example, Indigenous farmers use little tilling—no-till farming retains microbes and insects needed for healthy soil, increases water and nutrient absorption, slows evaporation, decreases erosion, reduces dependence on tractor fossil fuels, and sequesters carbon, thereby diminishing greenhouse gas emissions.[17] Indigenous farmers have also practiced agroforestry, combining crops and forests to create a cooperative ecosystem that stores carbon and protects yields from wind, rain, hail, and extreme temperatures. Crop rotation and polycultures reduce erosion and preserve soil quality, averting the need for pesticides, fungicides, herbicides, and fossil fuel–derived fertilizers.[18] Allocating a substantial portion of the hefty resources the US invests in animal agriculture (see "Of Cash Cows and Golden Eggs FACTS") into further research in this direction would go a long way toward solving numerous ethical and environmental problems.

Like my grandmother's pearls, I hold knowledge of Indigenous cultures close to my heart. Until well into adulthood, all the adapting and re-inventing necessary when I was a kid fragmented my personal history into jumbled heaps of snapshot and video clip memories. My ancestral roots help bind my splintered identity into a coherent whole. Embracing Indigeneity reaches far beyond my own personal benefit. Our current environmental crisis and racial reckoning cry out for a kind of profound individual and systemic code switching that rejects destructive colonial beliefs and habits.

Riffing on a Mexican *dicho* (saying): *Quisieron enterrarnos, pero no sabían que las plantas mas fuertes crecen de raíces nativas.* (They tried to bury us, but they did not know that the strongest plants grow from native roots.)

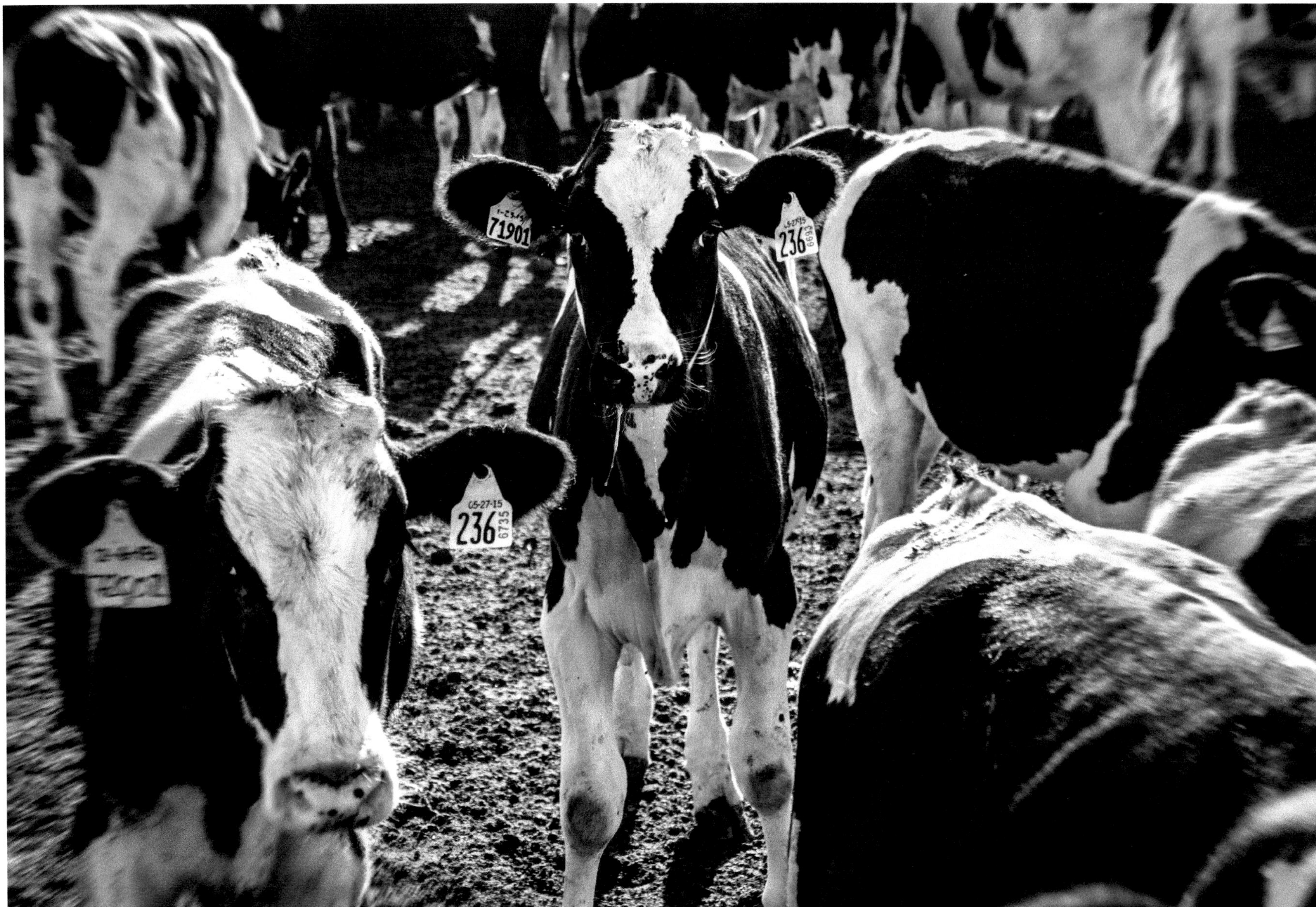

Calf 71901—Feedlot, 111 Degrees Fahrenheit, 2015

Feedlot, 111 Degrees Fahrenheit, 2015
38,500

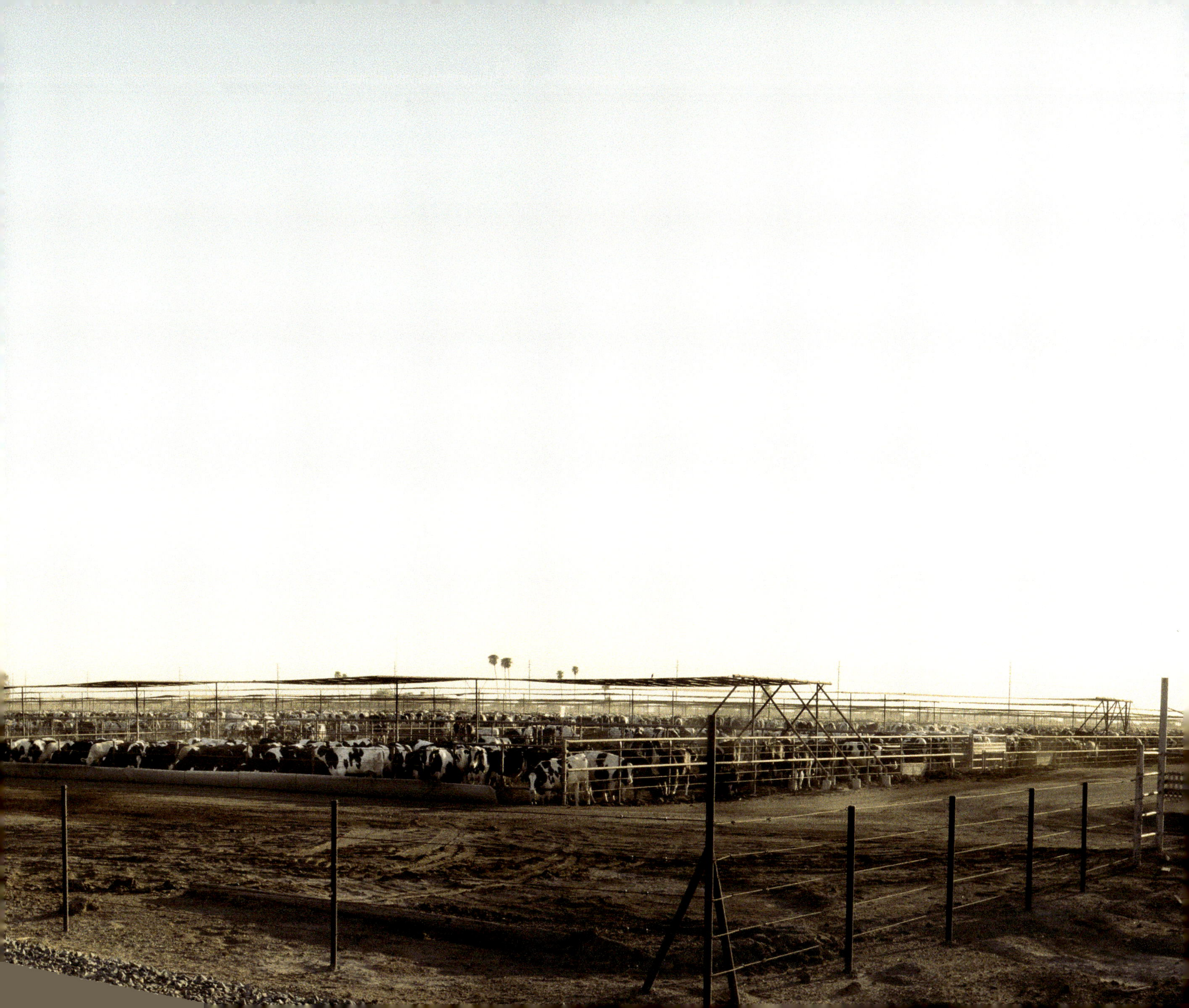

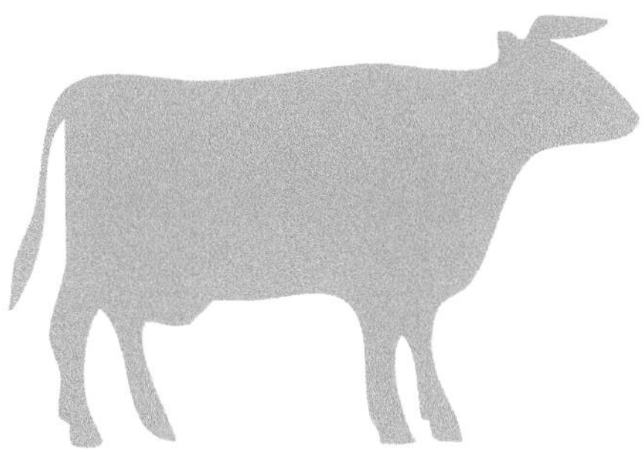

Photographing Feedlots

That day the heat blasting down from the sky and radiating up from the ground distorted our vision. By mid-afternoon, the temperature hit 116 degrees Fahrenheit. After less than half an hour out of my air-conditioned car, my face was brick red, and I felt light-headed.

In the evening, dirty and hungry, we arrived at a restaurant minutes before closing. Nevertheless, the waiter welcomed us. He was tall, muscular, friendly, and flirty. When we told him we're vegan, he brought us a treat: tofu kebabs that had been marinated in a savory sauce. They were delicious.

Then he announced that *he* was not vegan, he would *never* be vegan, he *loves* bacon.

I asked if it didn't bother him, about the pigs.

He said, "No, of course not. Animals don't have feelings."

We exclaimed that animals do have feelings. I would have said, we are all animals, humans are animals, but he didn't give me a chance.

He insisted animals don't have feelings.

We asked, "What about dogs? Do you have a dog?"

"Maybe dogs might have feelings. But the animals, they don't know they're going to die so it's okay to kill them. Maybe they're scared but it's not because they're scared of death. They're just scared."

I pointed out that fear is a feeling.

He agreed that fear is a feeling and repeated that he would never stop eating meat, that eating meat was the reason for his handsome body.

We did not pursue the conversation much further. I didn't want to waste time in a tiresome discussion. I simply wanted to enjoy dinner with my dear friend.

Afterward, we spent the night at a hotel with points I'd racked up on my credit card. It was plush in the way that's popular among the not quite ultra-rich. The lawns were perfectly manicured. In the desert, during the fourth year of the California drought, the grass was watered daily. There were duck ponds, and three swimming pools. The lush terrain presented a freaky contrast to the acres of barren, excrement-covered dirt on which cattle stand, or lie down, all day and night, day after day, the ammonia fumes searing their eyes, nostrils, mouths, and tongues, with no real shade, only water from a dirty concrete tub to drink, no respite until the abject night from that hell of over one hundred degrees Fahrenheit.

The industry name for the places I photographed is CAFO—Confined Animal Feeding Operation. Most people call them factory farms. Ninety-nine percent of meat, dairy, and eggs comes from animals on factory farms.[1] Feedlots confine cattle by the thousands for the beef or dairy industries. The animals are young, most likely less than a year old. They are fed an unnatural grain diet to fatten them as quickly as possible,[2] which can cause gastric distress. [3] They have been painfully dehorned, branded, and/or castrated, with no requirements for anesthesia.[4]

Comparisons have been made to the Nazi concentration camps. Isaac Basevich Singer wrote, "In relation to them all people are Nazis; for the animals, it is an eternal Treblinka."[5] Alex Hershaft, Holocaust survivor and founder of Farm Animal Rights Movement, said, "Millions knew about the death camps in their midst but pretended not to notice—just as we pretend not to notice factory farms, slaughterhouses and factories in our neighborhoods."[6] Some people jokingly call these places "Cowschwitz." Of course, comparisons only go so far; every oppression has a quality and history all its own.

I'm sure in that hotel people were suffering—couples at odds with their spouses, on the verge of divorce or going through a divorce, maybe frustrated with their job or career, perhaps their child had drug problems, or a parent was dying. The ways in which humans suffer are endless. Suffering cannot be quantified.

I'm also sure no one in that hotel would ever trade places with any one of the animals we wept for that day. And every one of those animals would give anything to spend just a short time in that luxury resort.

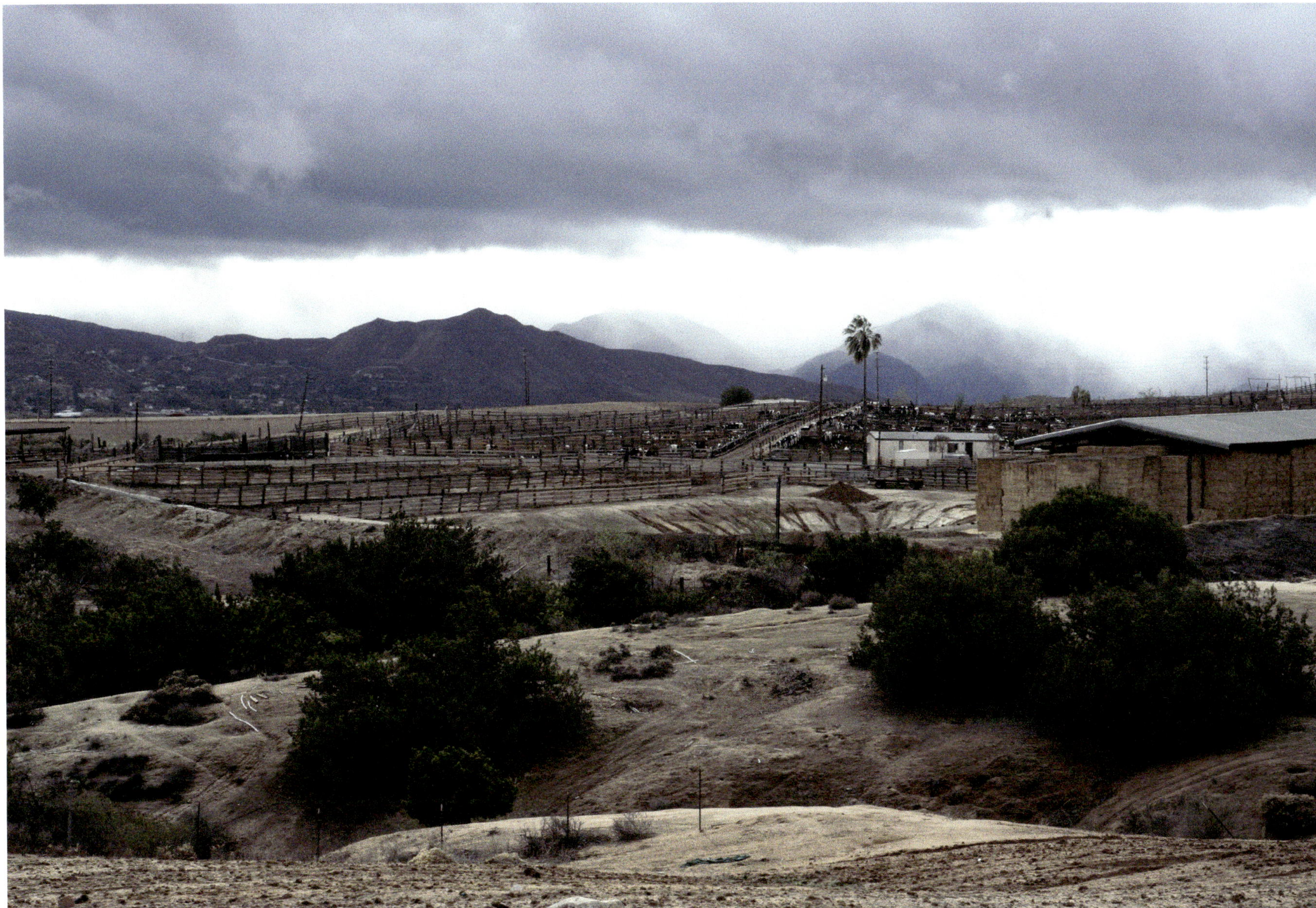

Dairy Farm (Cows), 2015
3,000

EXCERPT FROM

The Eighth Duino Elegy

Rainer Maria Rilke (1875-1926)

If there was consciousness like ours
in the sure creature, that moves towards us
on a different track—it would drag us
round in its wake. But its own being
is boundless, unfathomable, and without a view
of its condition, pure as its outward gaze.
And where we see future it sees everything,
and itself in everything, and is healed for ever.[1]

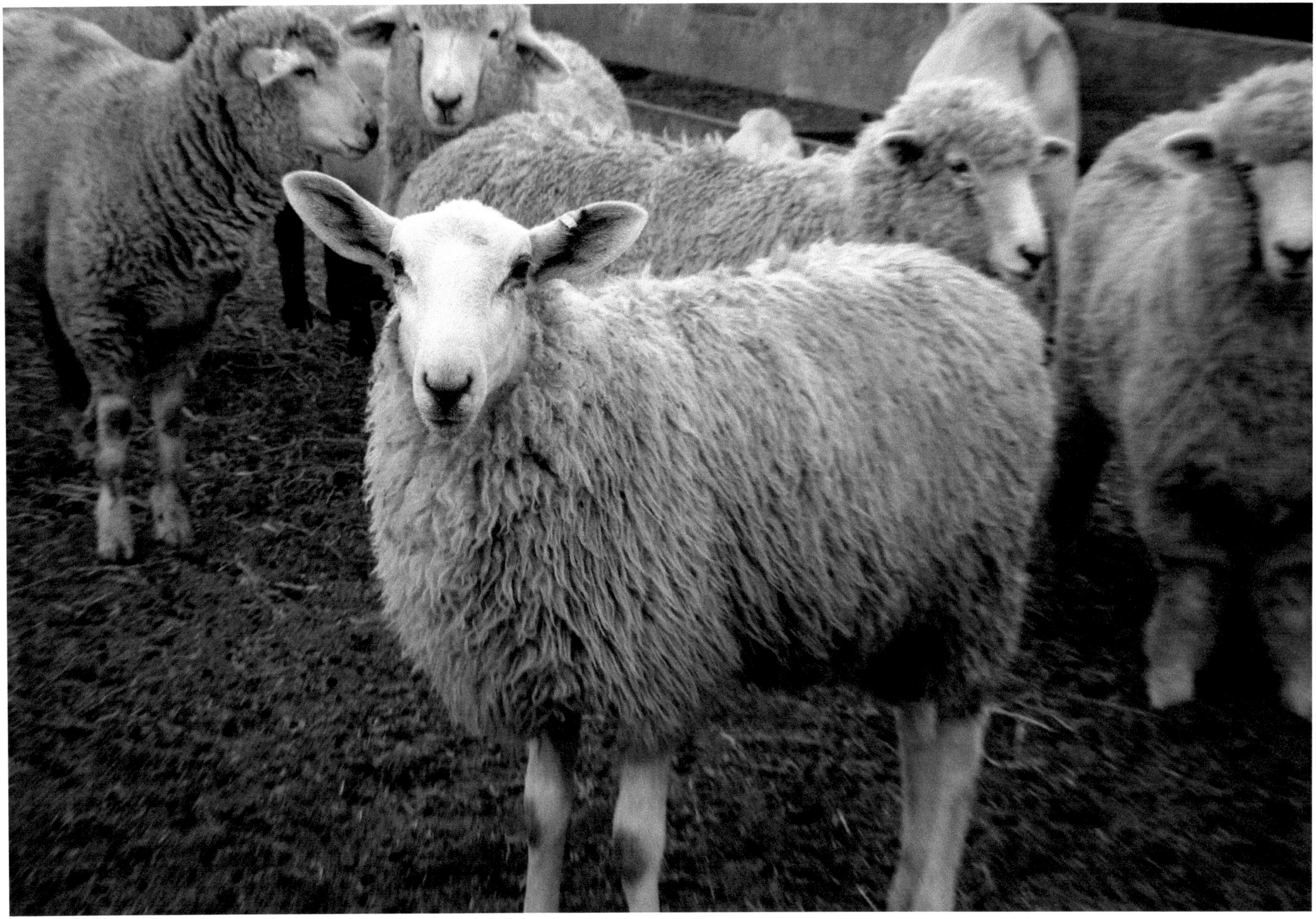

Sheep—Livestock Auction, 2015

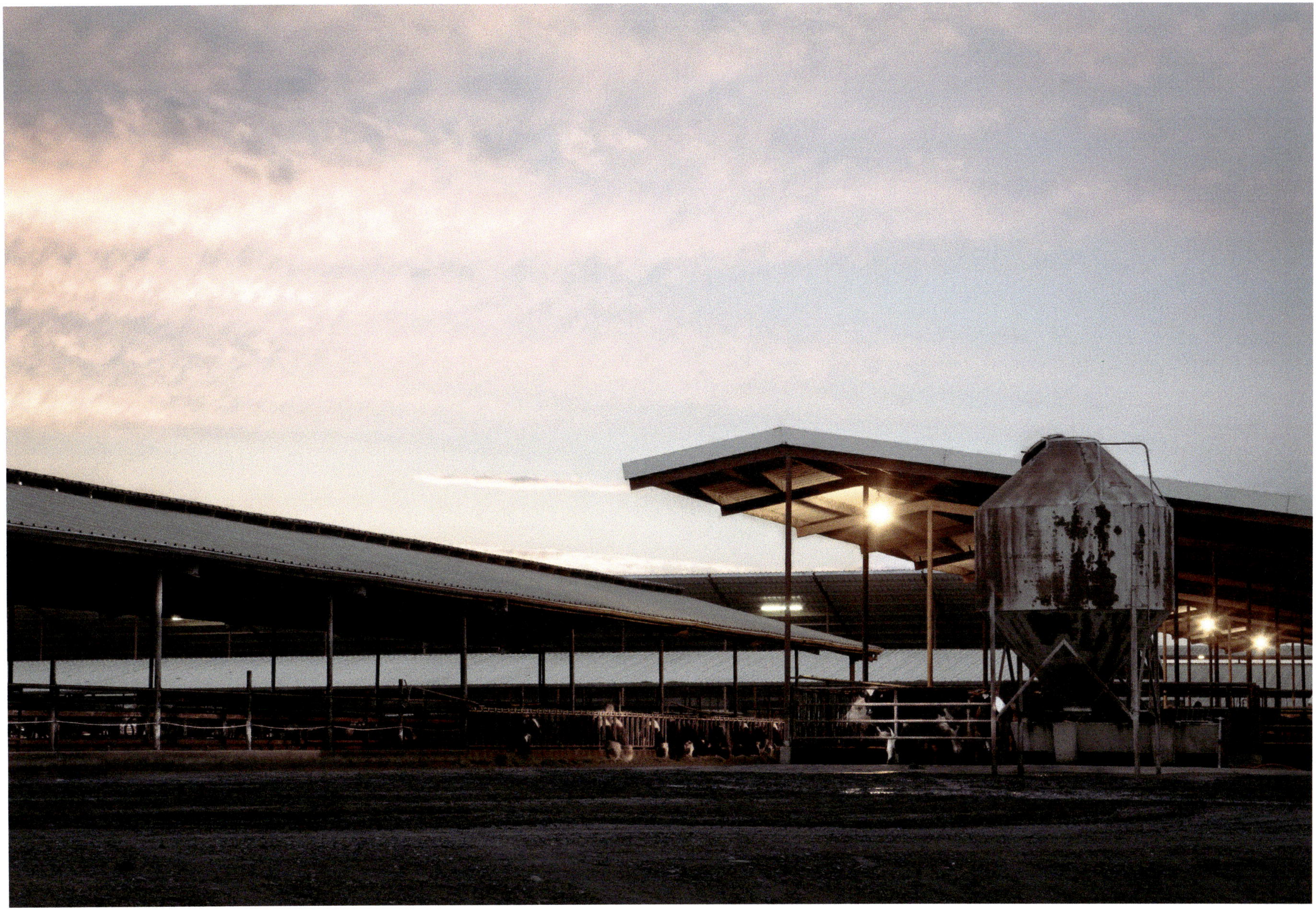

Dairy Farm (Cows), 2016
3,400

Environmental Impacts

Alarms are clanging louder than thunder and flashing brighter than lightning. As of April 23, 2023, CO_2 levels climbed to 423.23 parts per million (ppm), up over four ppm from the previous year.[1] Environmental scientists have warned that the highest safe level of carbon dioxide in the atmosphere is 350 ppm.[2] Elevated levels of greenhouse gases are to blame for global warming, loss of glaciers and sea ice, drought, wildfires, heat waves, sea-level rise, severe storms, flooding, water and food insecurity, habitat loss,[3] and mass species extinction.[4] Our current environmental emergency has been caused by human activity and disproportionately affects the most vulnerable—people of color, the economically disadvantaged, and nonhuman animals.[5]

- Though it has been estimated to be as high as 51 percent,[6] the most reliable study estimates that livestock production is responsible for 18 percent of global greenhouse gas emissions, more than all global transportation, including trains, planes, cars, and marine vessels,[7] and twice those of plant-based foods.[8]
- Raising animals for food produces 32 percent of anthropogenic methane in the US. Methane is eighty times more powerful over twenty years at warming the Earth than carbon dioxide.[9]
- Impacts of the lowest-impact animal products exceed average impacts of substitute vegetable proteins across greenhouse gas emissions, eutrophication, and acidification.[10]
- Raising animals for food requires almost 83 percent of global agricultural land but produces only 37 percent of the edible protein and only 18 percent of the world's supply of calories.[11]
- From 2000 to 2015, deforestation caused by cattle ranching accounted for 16 percent of deforestation. This did not include deforestation due to soy crops fed to livestock, which is estimated to account for at least 8 percent of deforestation.[12] Deforestation contributes to climate change, air pollution, pandemics, soil degradation, sea-level rise, loss of biodiversity, habitat destruction, and species extinction.[13]
- In 2013, 48.7 percent of corn grown in the US[14] and over 70 percent of soybeans was fed to farmed animals.[15] These two crops account for more than half of US crops.[16] As intensified agricultural production has increased in the twentieth century, 75 percent of the world's crop diversity has been lost.[17]
- Lack of biodiversity reduces soil fertility by depriving it of essential bacteria, fungi, earthworms, nematodes, insects, and other invertebrates.[18] Loss of biodiversity also degrades soil structure, which results in soil erosion.[19]
- Large amounts of synthetic fertilizers used to boost soil fertility and increase crop yields are responsible for more greenhouse gas emissions than commercial avaiation.[20]
- Herbicide use by 1980 climbed to cover 90–99 percent of US corn and soybeans.[21] These crops also depend heavily on pesticides. Herbicides and pesticides are harmful to wildlife and pollinators. The chemicals leave residue in the soil, polluting groundwater and waterways, which damages ecosystems, further decreasing biodiversity.[22]

- International affairs think tank, Chatham House, lists change to predominately plant-based diets as the first of three key levers in support of biodiversity and improved public health.[23]

- Producing animal-derived products requires more water than the equivalent nutritional value of plant products. Per gram of protein, milk, eggs, and chicken meat production requires one and a half times the amount of water than for pulses such as beans and lentils. Producing beef requires six times the amount of water per gram of protein than for pulses and twenty times more water than for cereals and starchy roots.[24]

- One-third of the world's population faces high water stress. In the US, five states experience high water stress and an additional nine states face medium to high water stress.[25]

- Water shortages are to blame for increased death, conflicts, and migration.[26]

- As of 2010, US Animal Feeding Operations (AFOs, also known as factory farms, from which 99 percent of meat, dairy, and eggs come[27]) generate forty times more raw waste than is generated from wastewater treatment plants. However, as of 2022, farmed animal waste is unregulated and treatment is not required by the US federal government.[28]

- Waste on industrialized animal farms is most often stored in lagoons[29] that also contain pesticides and other chemicals, antibiotic residue, and even dead animals.[30] Spills are often observed. The slurry is also applied to crop fields. The excess waste runs off the land, pollutes waterways,[31] and creates algal blooms[32] and dead zones. Aquatic life cannot survive in dead zones.[33]

- CAFOs (a subset of AFOs) expose animals, workers, and neighboring communities to elevated levels of hydrogen sulfide, ammonia, and particulate matter.[34]

- Farmed fish feed as well as farmed fish waste result in excess nitrogen and phosphorous (eutrophication) in marine ecosystems. Eutrophication is characterized by algal blooms, dead zones (hypoxia or anoxia), and ocean acidification, which produce toxins that affect human and animal health.[35]

- Escaped farmed fish, which number in the hundreds of thousands and in some cases in the millions, spread deadly diseases and parasites to wild fish.

- Escaped farmed fish prey on young wild fish and compete with native marine populations for food, habitat, and other resources.
- Interbreeding between farmed and wild salmon results in loss of biodiversity and inability to adapt to environmental changes.[36]
- Escaped farmed fish establish themselves as invasive species; for example, thousands of farmed Atlantic salmon that escaped in Pacific waters off Washington's Cypress Islands in 2017.[37]
- Wild fish fed to farmed fish may be contaminated with PCBs, dioxins, and other poisons, which accumulate in their bodies and are then consumed by humans.[38]
- Antibiotics and pesticides are heavily used in farming fish. Antibiotics and pesticides contaminate waterways and wild marine life.[39]
- Wildlife Services, an arm of the USDA, exterminates wild animals primarily for the benefit of private ranchers.[40] In 2021, Wildlife Services killed 404,538 native animals, including gray wolves, coyotes, black bears, mountain lions, bobcats, foxes, beavers, eagles, falcons, hawks, and owls. They also destroyed coyote dens as well as pocket gopher and prairie dog burrows. This number was relatively low; in 2019, Wildlife Services killed 1.2 million native animals, and they've killed as many as five million animals a year.[41] A 2016 investigation by Christopher Ketcham of *Harper's Magazine* reported, "Wildlife Services gunned down coyotes from airplanes and helicopters. Its trappers used poison baits, cyanide traps, leghold traps, and neck snares. They hauled coyote pups from dens with lengths of barbed wire, strangled them, or clubbed them. Sometimes they set the animals on fire in their dens or suffocated them with explosive cartridges of carbon monoxide." Ketcham's sources were former Wildlife Services agents who have witnessed or participated in these practices themselves.[42]
- A 2022 study shows that "rapid global phaseout of animal agriculture has the potential to stabilize greenhouse gas levels for thirty years and offset 68 percent of CO_2 emissions this century."[43]
- Numerous studies show the environmental benefits of a plant-based diet.[44]

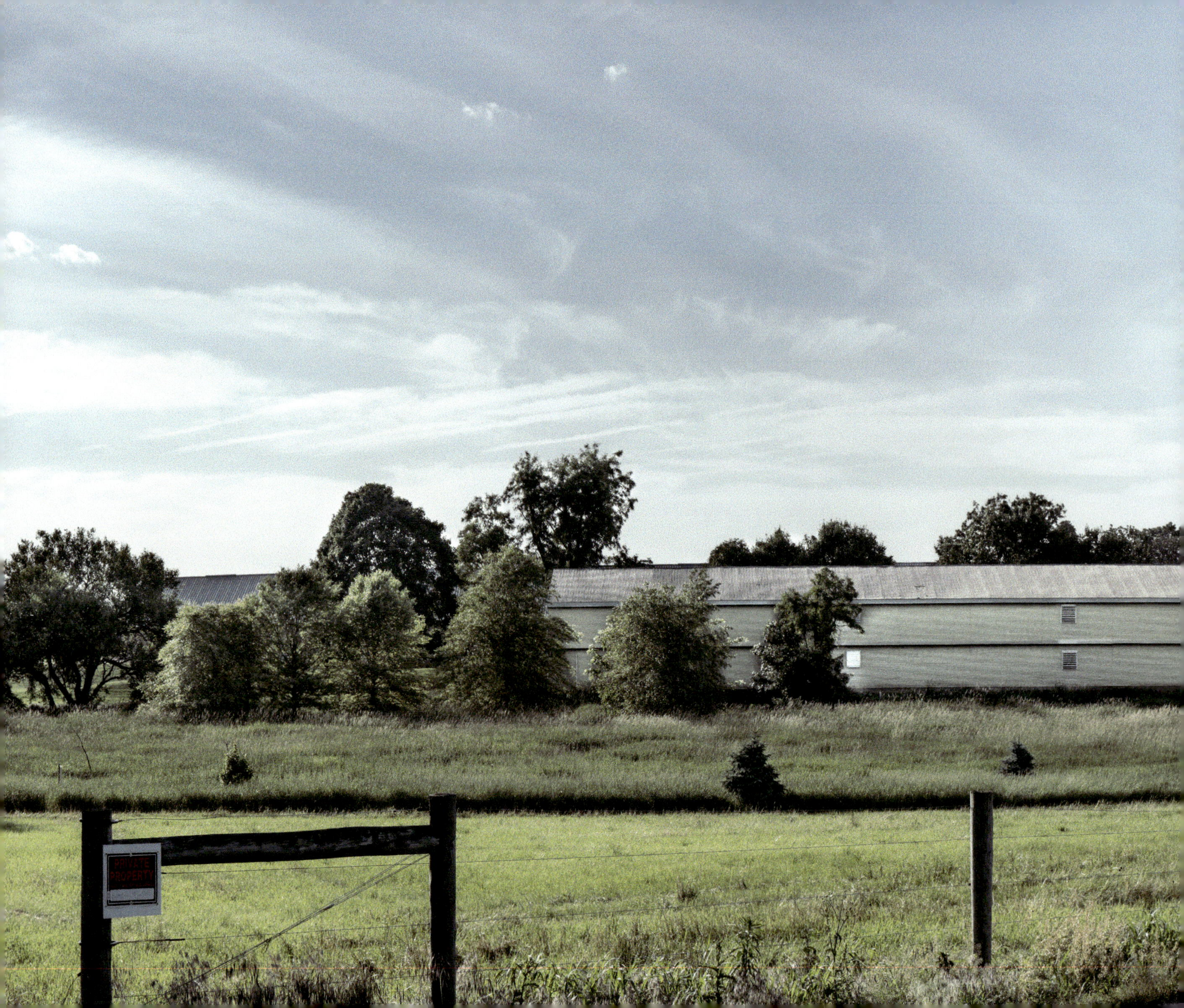
PRIVATE
PROPERTY

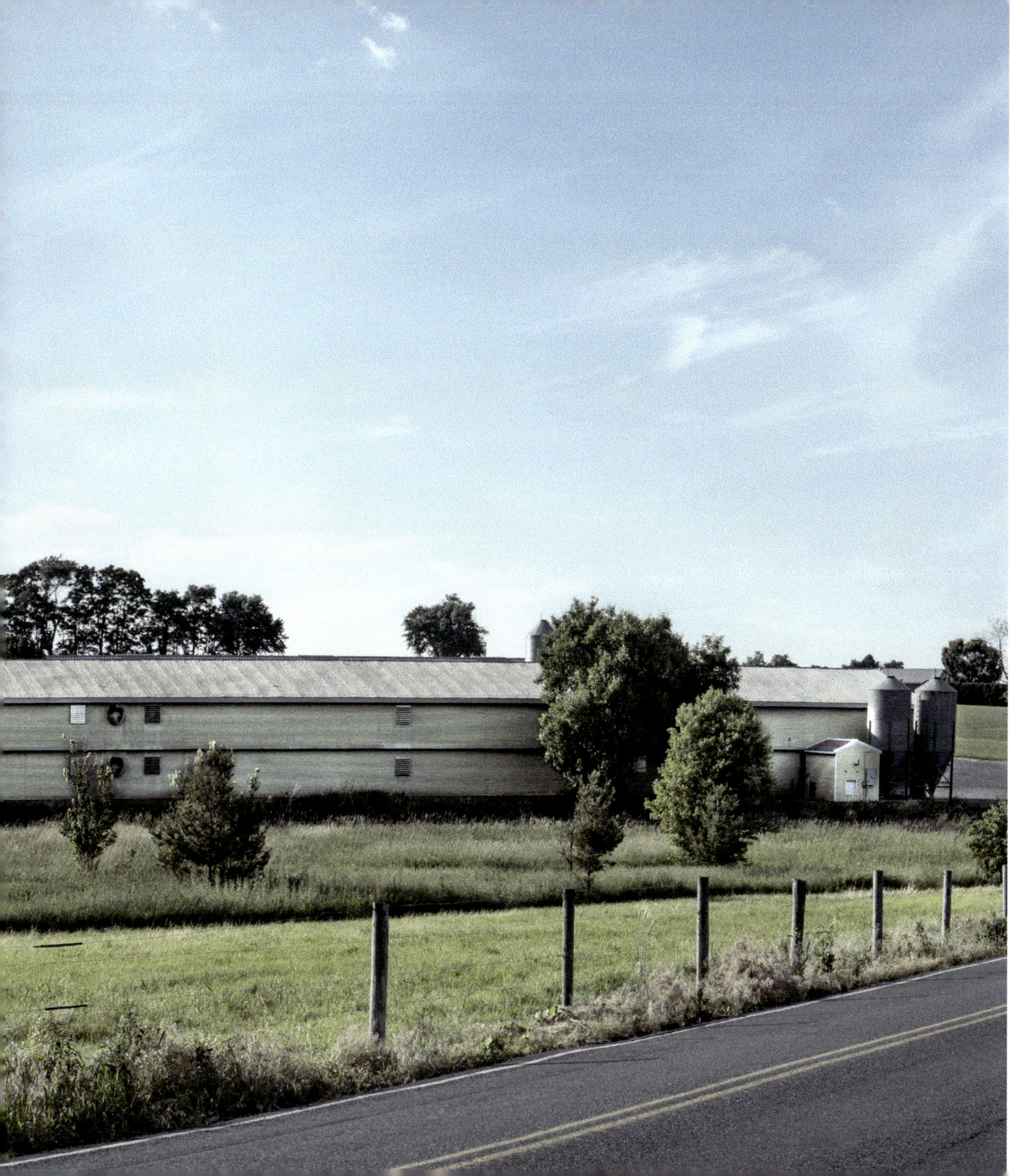

Organic Chicken Farm, 2016
72,200

Dairy Pride

Chloe Hanson

Like any TV father, I'm sure
he calls his daughter *princess*

while serving her a glass of milk,
white, thick, unblemished

as her skin, both mother-fresh,
newly-poured. Americana poster-children,

this family draws water from private wells.
Meanwhile the surface is no longer safe

for swimming, drinking. But there's always
milk in the fridge. Glass-bottled nostalgia.

Father turns to camera. *Do you know*
he asks, *dairy is both healthy and sustainable?*

He tousles TV-daughter's hair, she sips,
smiles with wet upper lip. *Thanks, Dad!*

She's not his child any more than the penned
creature in the next shot is her mother,

no matter how this girl suckles
at the swollen teat. The father pours

another glass, this one for show only:
glue thinned with water, impossible—

white.

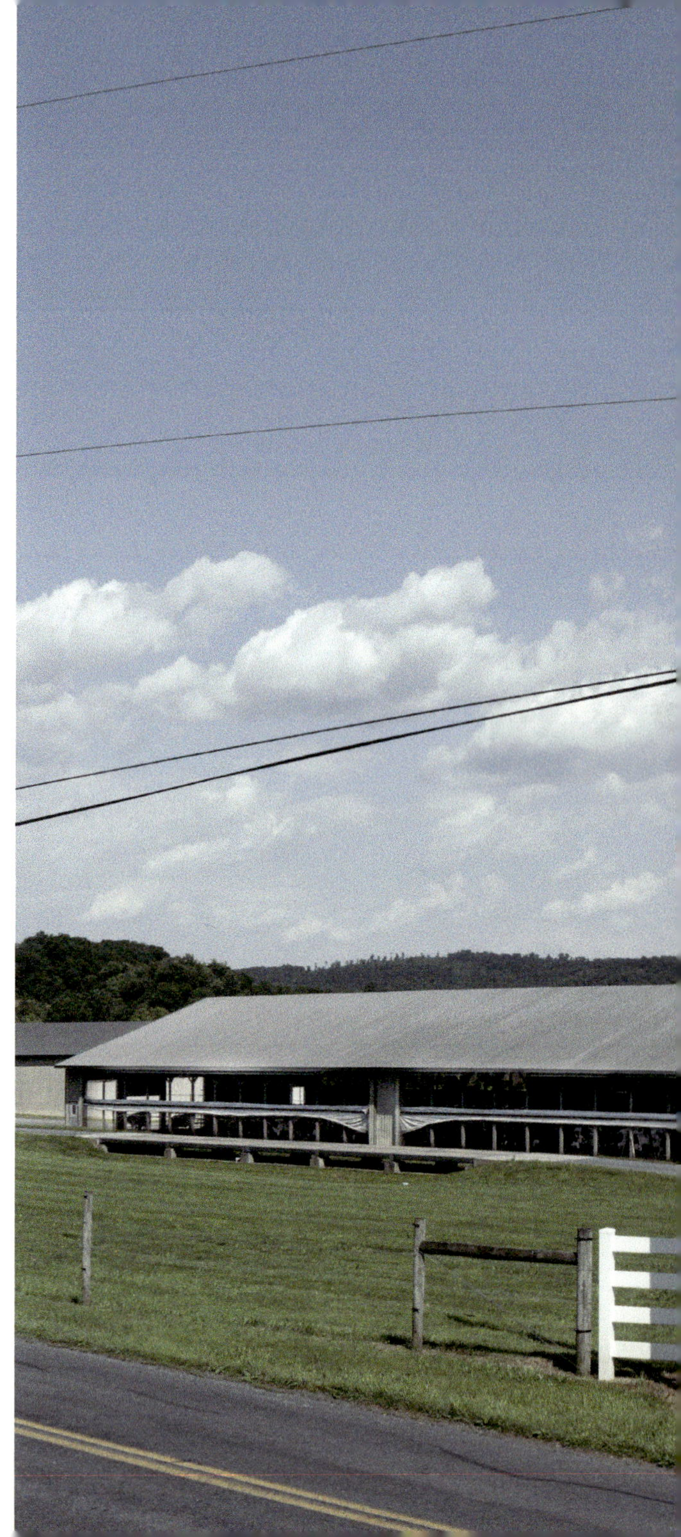

Dairy Farm 2 (Cows), 2016
1,150

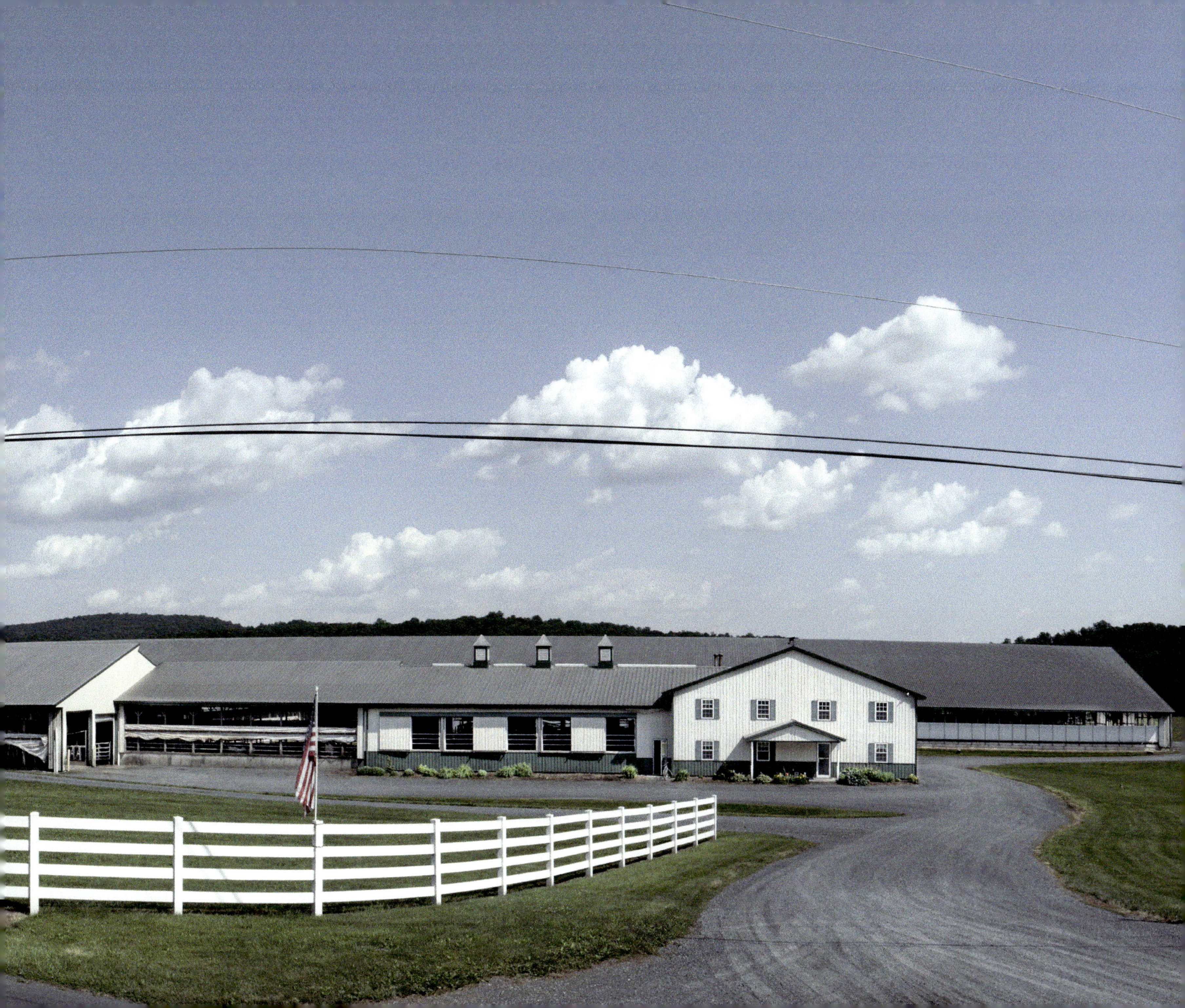

EXCERPT FROM

The Jungle

Upton Sinclair (1878–1968)

There was a long line of hogs, with squeals and lifeblood ebbing away together; until at last each started again, and vanished with a splash into a huge vat of boiling water.

It was all so very businesslike that one watched it fascinated. It was pork-making by machinery, pork-making by applied mathematics. And yet somehow the most matter-of-fact person could not help thinking of the hogs; they were so innocent, they came so very trustingly; and they were so very human in their protests—and so perfectly within their rights! They had done nothing to deserve it; and it was adding insult to injury, as the thing was done here, swinging them up in this cold-blooded, impersonal way, without a pretense of apology, without the homage of a tear. Now and then a visitor wept, to be sure; but this slaughtering machine ran on, visitors or no visitors. It was like some horrible crime committed in a dungeon, all unseen and unheeded, buried out of sight and of memory.

One could not stand and watch very long without becoming philosophical, without beginning to deal in symbols and similes, and to hear the hog squeal of the universe. Was it permitted to believe that there was nowhere upon the earth, or above the earth, a heaven for hogs, where they were requited for all this suffering? Each one of these hogs was a separate creature. Some were white hogs, some were black; some were brown, some were spotted; some were old, some young; some were long and lean, some were monstrous. And each of them had an individuality of his own, a will of his own, a hope and a heart's desire; each was full of self-confidence, of self-importance, and a sense of dignity.[1]

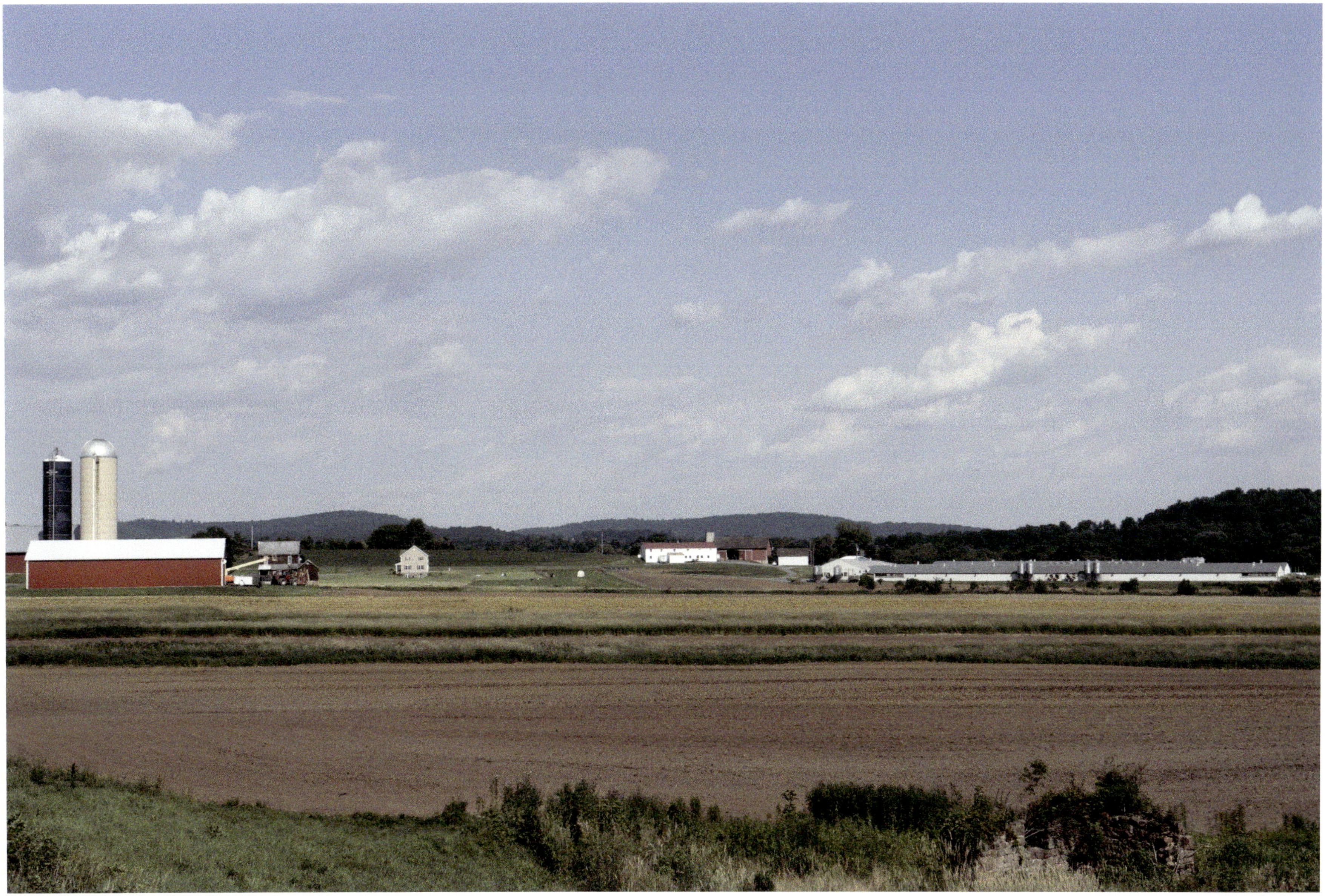

Pig Farm, 2016
8,000

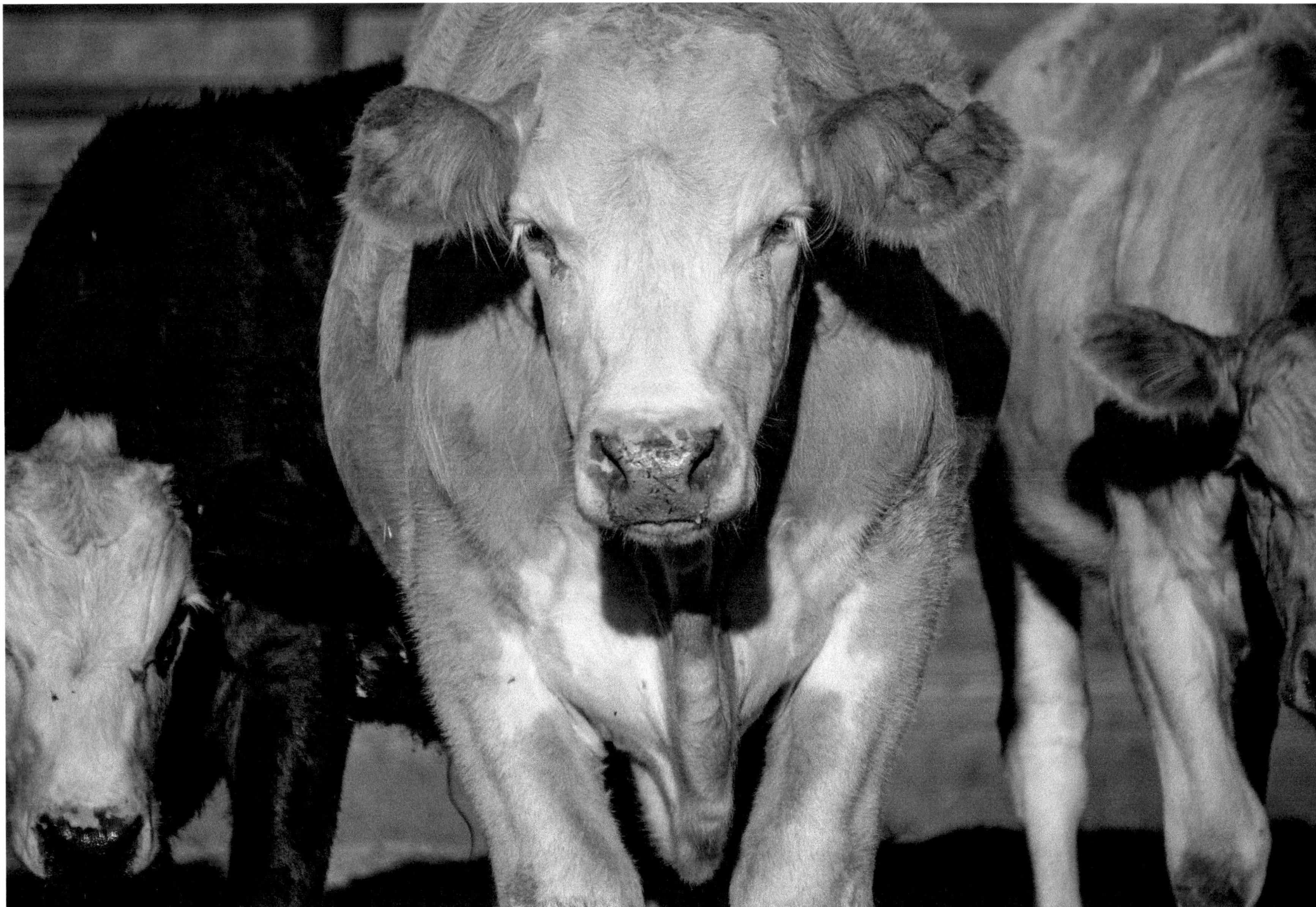

Golden—Grass-Fed Beef Ranch, 2014

Grass-Fed Greenwash

Imagine you are out for a drive on a cloud-free Sunday afternoon. Perhaps you're on your way to the farmer's market or to visit a beloved friend—someplace you've been looking forward to. At the crest of a gentle grade, the land rolls out before you as in a fairy tale: grass-fed cattle serenely grazing on rolling hills carpeted in a zillion shades of rain-charged green.

"Grass-fed" commonly refers to cattle raised for beef who are fed grass and forage their entire lives. A jewel in the crown of the slow food, farm-to-table, locavore movement, grass-fed beef is the darling of meat-loving progressives all over the US and beyond. A nostalgic back-to-the-land ethos, before the evils of industrialization, when you knew where your food came from, forms the basis for the movement's allure. Proponents of the movement claim to hark back to a better, simpler past—the progressive equivalent of Make America Great Again. Pointing to the animals' ability to roam relatively free, grass-fed beef advocates boast that the animals are treated better than animals in industrial facilities, which makes their meat tastier.

Grass-fed beef is purported to be environmentally friendly! The environmental appeal of "local meat" seems obvious: local economies benefit and less greenhouse gas is emitted in the transport. Grass-fed beef ranchers contend that the domesticated cattle they raise mimic the grazing patterns of ancient ruminants, therefore their products are more sustainable. Their claims are primarily based on a 2013 Ted Talk by cattle rancher Allan Savory. Savory argued that the trampling of moving herds of large animals encourages the soil to retain water and regrow carbon-sequestering grasslands, thus reducing greenhouse gases.[1]

But like a fairy tale, the narrative of grass-fed beef is a fiction. The US government does not regulate the use of the term "grass-fed."[2] The difference between raising grass-fed cattle and raising cattle by the feedlot method is only a matter of how the animals spend the last five months of their lives on average. All cattle raised for beef graze on forage and grass until they are up to about a year old. The great majority of these animals are then sent to feedlots where they are "finished" on grain-based feed to hasten weight gain until they are slaughtered when they are about fifteen to twenty-eight months old.[3] Only about 5 percent of cattle raised for beef are "grass-fed."[4] The slaughtering methods are the same for grass-fed cattle as for cattle finished on feedlots[5] (see "Bad Boss").

Finishing cattle on pasture is even less sustainable than grain-finished cattle. Grass-fed beef requires much more land. Cattle farmers who finish their cattle on pasture must rotate their herds so the fields can recover between grazing.

The land must be aggressively managed and can't be used for any other purpose.[6] Producing the same amount of grass-fed beef as all the beef currently sold in the US would require more than three times the land.[7] Because animals raised on pasture take longer to reach slaughter weight, they consume more water and other resources.[8] The production of grass-fed beef also creates more greenhouse gases;[9] because they live longer, grass-fed cattle emit more methane,[10] a greenhouse gas twenty-five times as potent as carbon dioxide.[11] Allan Savory's claim that raising cattle on pasture reduces greenhouse gases has been roundly debunked.[12] Savory is also responsible for the extermination of 40,000 elephants in Zimbabwe, allegedly to prevent desertification, only to make it worse.[13] And grass-fed meat is not necessarily local—about 75 percent to 80 percent of grass-fed beef sold in the US comes from Australia, New Zealand, and South America.[14] "Buying local" is a much less effective means of lowering an average household's climate footprint than a plant-based diet.[15]

The idealized view of a pastoral past is the opposite of progressive. Indigenous peoples in pre-Columbian North and Central America did not farm animals. The European settler colonialists brought cows, pigs, sheep, goats, and chickens to the Americas,[16] along with a legacy of white supremacy and ownership, exploitation, and pollution of the land. Raising animals for food is how we got into the industrialized agriculture horror show we're stuck in; twentieth-century farmers who raised animals on pasture were forced to increase their productivity to meet the market demand at a cheap price, thus factory farming.[17] Grass-fed meat costs more and is therefore unavailable to many Americans; in 2015, grass-fed beef cost an average of $2.50 more per pound.[18] Most cattle farmers do not have the land or resources to raise cows on pasture until slaughter at a competitive price.

The glittering promise of environmentally superior beef is but an expensive, elitist fraud contrived from the bodies of animals who have nothing else to steal.

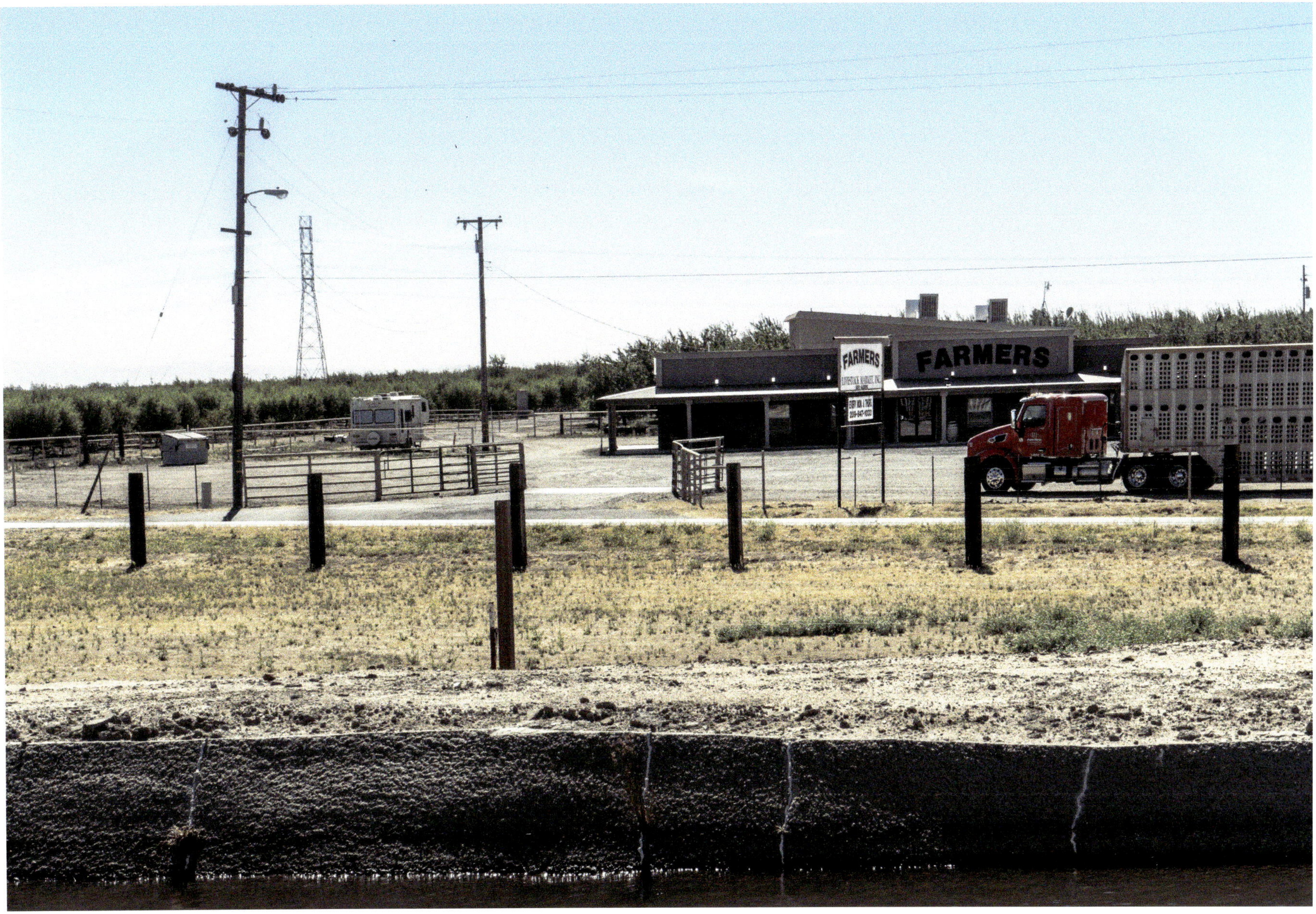

Livestock Auction, 2016
"proprietary"

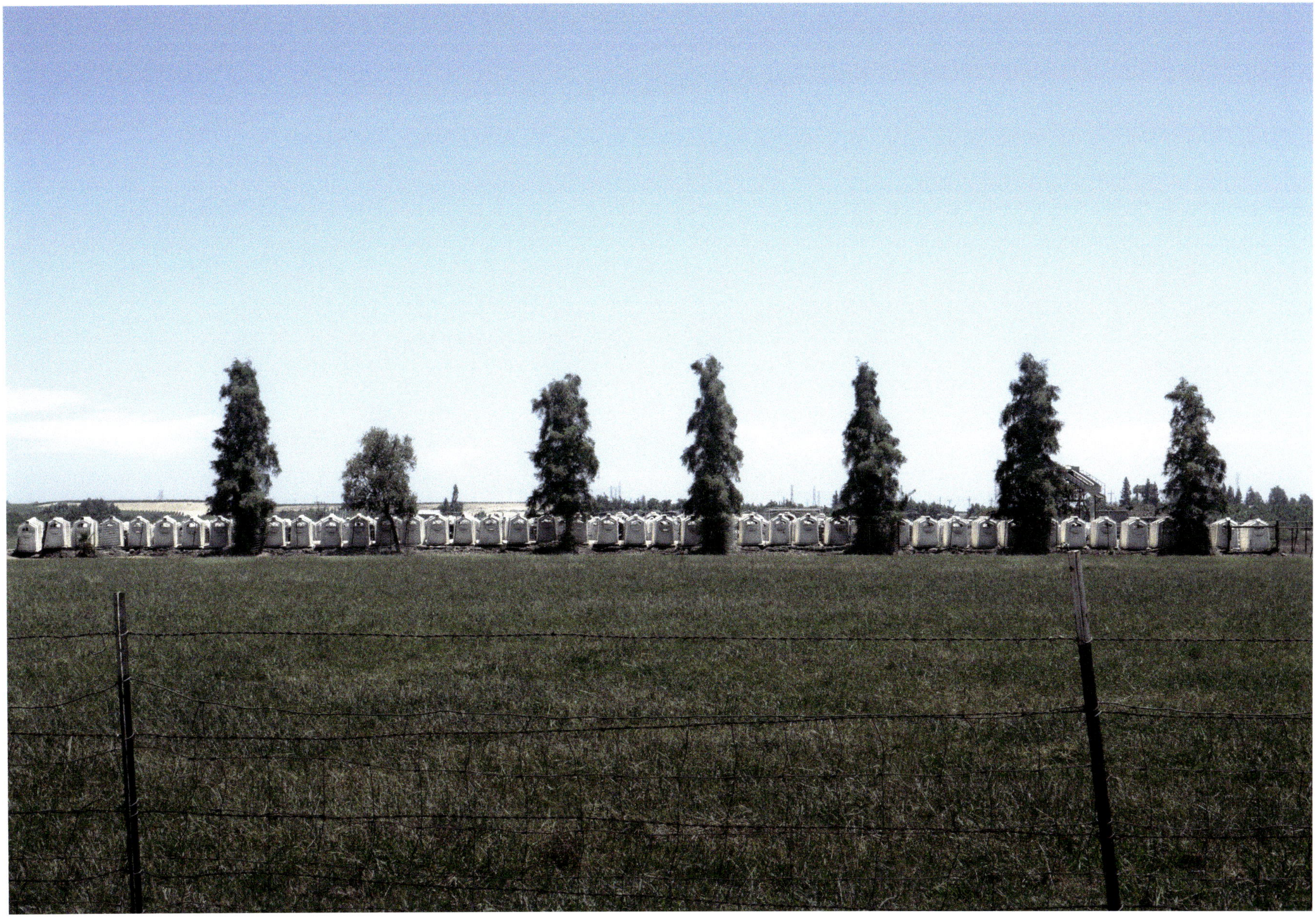

Dairy Calf Hutches, 2016
1,700

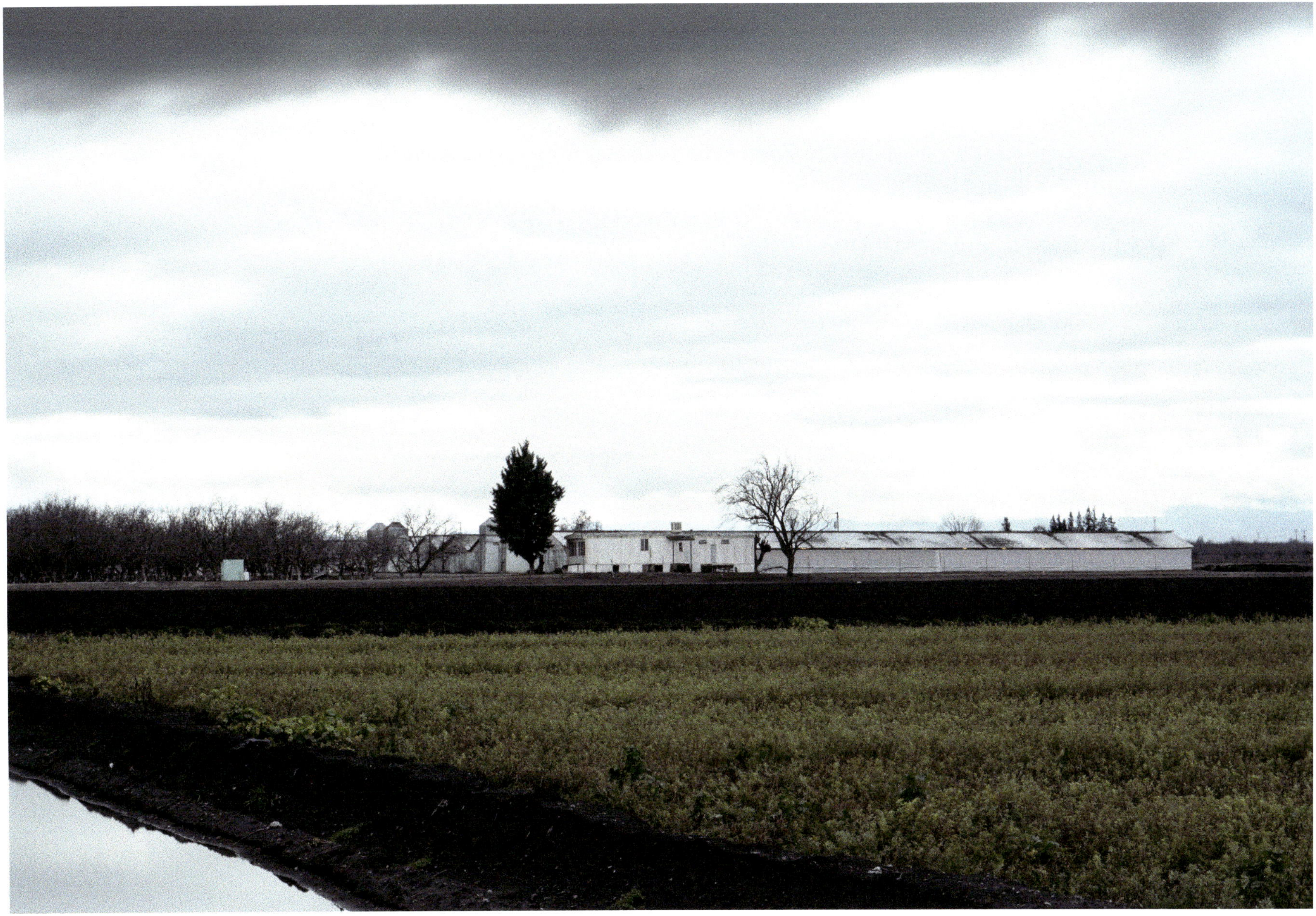

Egg Farm, 2016
180,000

Nonconsensual Sex

Many words came to mind as I researched this subject. I've decided that an unvarnished industry description speaks for itself.

Virtually all turkeys raised for food are bred using a process the industry calls artificial insemination.[1] A worker stimulates the male genitals by massaging the abdomen and the area over the testes. While pushing on the tail with one hand, he milks the sperm from the genitals with the thumb and forefinger of the other hand. To inseminate a hen, a worker massages her genitals until her oviduct protrudes. He then inserts a syringe containing sperm into her oviduct.[2]

Though this may seem obvious, many people are unaware that cows in the dairy industry must be pregnant to produce milk. As of 2011, 95 percent of cows bred for their milk in the US are kept pregnant using artificial insemination.[3] About 6 percent of cattle raised for beef in the US are also artificially inseminated as of 2020.[4] To inseminate a cow, a worker clears her vulva, inserts his arm into her rectum, and locates her cervix. He then inserts a semen gun into her cervix and deposits sperm into her uterus.[5] This process is repeated until she is pregnant and within months after each birth. When she no longer produces enough milk to be profitable, at about three to four years old, she is sent to slaughter (see "Bad Boss").[6] To collect sperm, a teaser animal, often a steer castrated and injected with female sex hormones, is used to sexually excite the bull. Then a worker places the bull's penis in an artificial vagina and collects his semen. An alternative method is electroejaculation. A worker massages the bull's genitals to sexually excite him. Then he inserts a probe into the bull's anus and releases a series of electrical pulses of one to two seconds each and of increasing voltage until he ejaculates.[7] This is a painful procedure and has been banned in some countries.[8]

Pigs raised for food are also bred using artificial insemination.[9] To collect semen, a boar stud is trained to mount a dummy sow. A worker firmly grabs his penis and exerts pressure on it to simulate a female pig's cervix until the boar ejaculates.[10] To inseminate a female pig, a worker clears her vulva then stimulates her genitals until she produces uterine contractions. He then inserts a breeding rod into her cervix and dispenses semen.[11] A female breeding pig is repeatedly subjected to this process to produce as many litters as possible. When she is no longer fertile enough to be profitable, she is sent to slaughter (see "Bad Boss").[12]

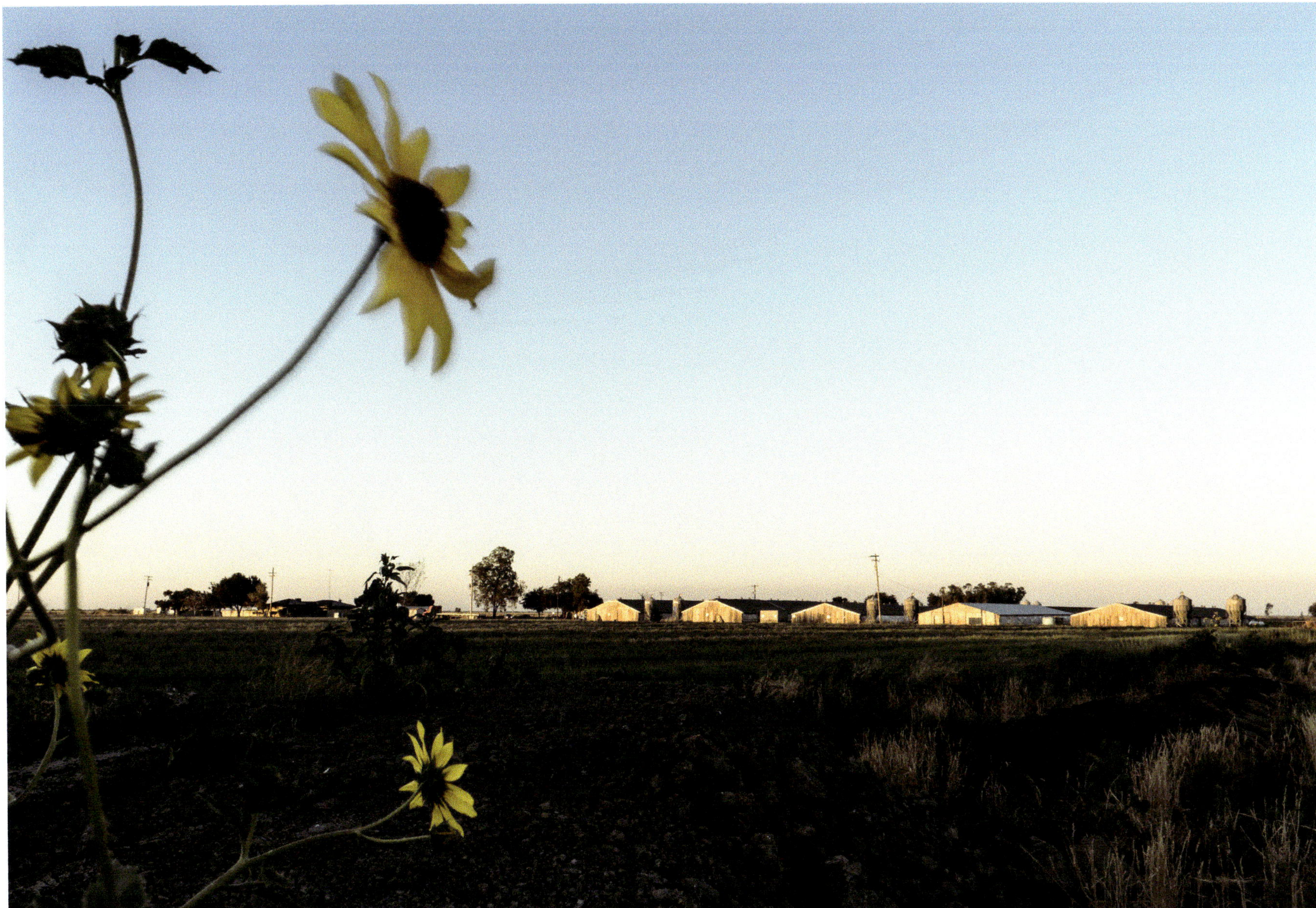

Turkey Farm, 2016
192,750

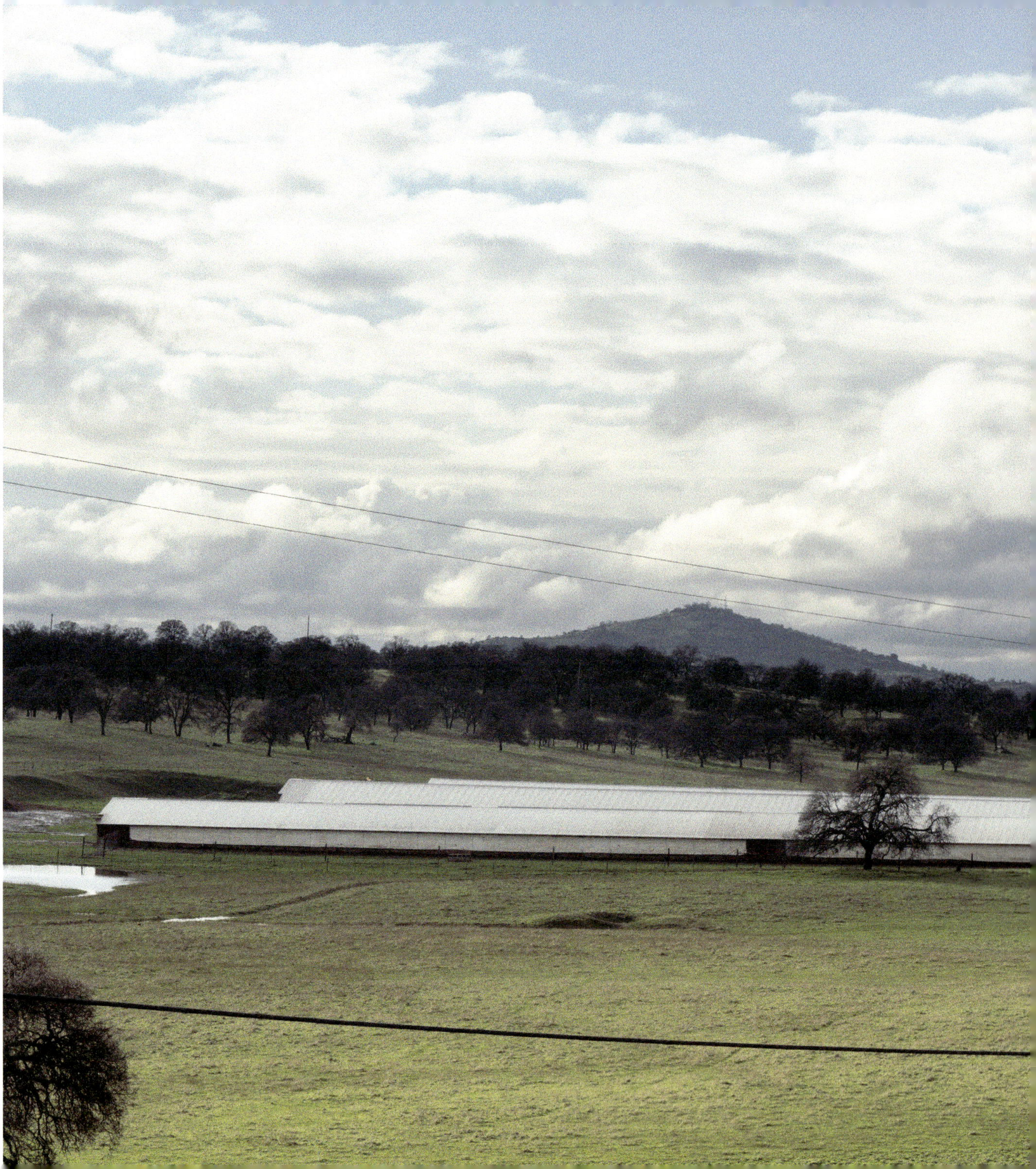

Family-Owned
Turkey Farm, 2017
310,000

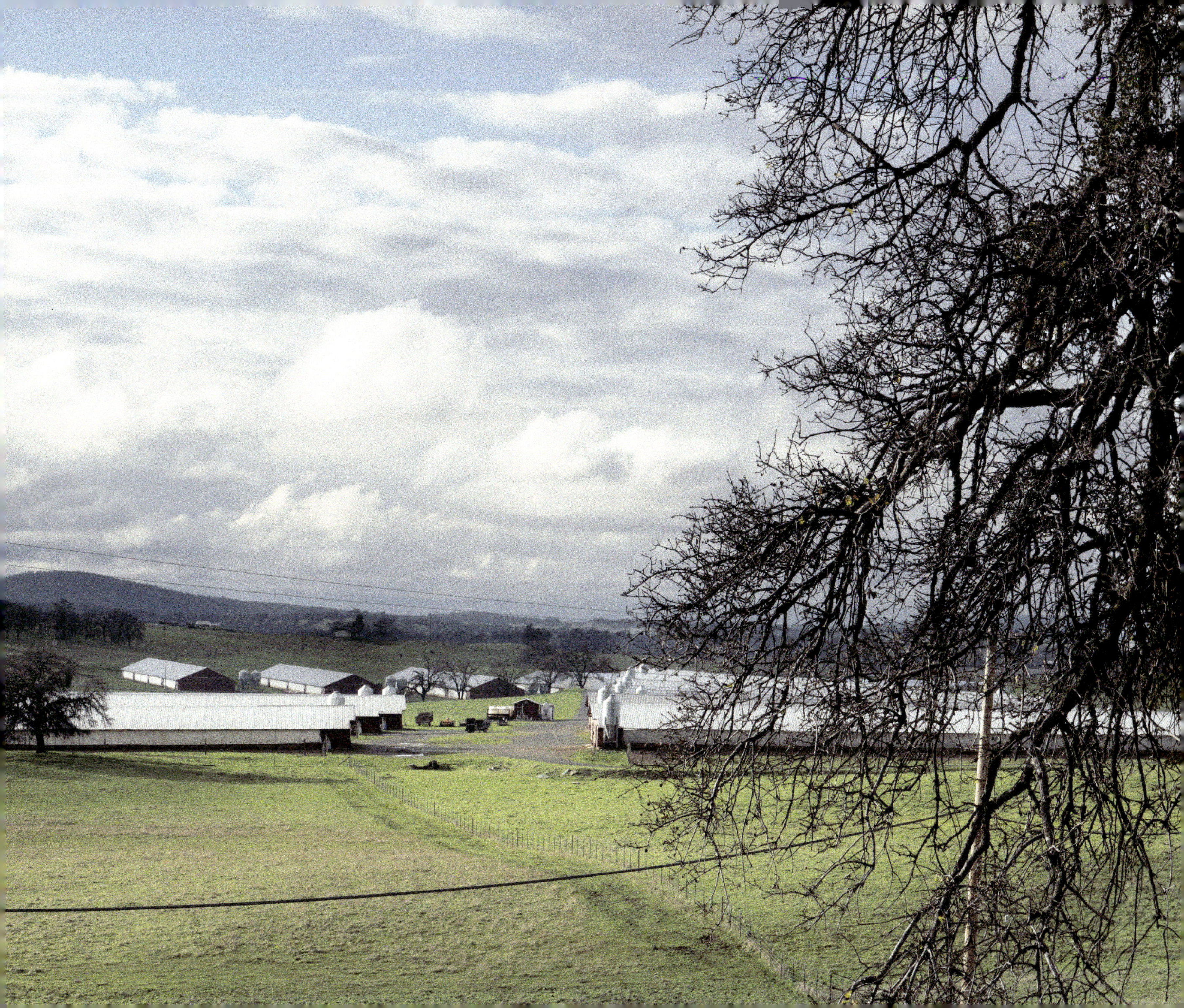

It's a Family Affair

The word "family" is saturated with significance and yet is so elemental as to obviate definition. Our most formative relationships are with family. Emotional ties to family are so deep they permeate our dreams. Family is foundational to identity and values. Lifelong warmth, comfort, and love are often associated with family.

But when applied to agriculture, the word is a betrayal. The families of nonhuman animals bred for food are torn apart in the service of profit. The USDA classifies a family farm as "any farm organized as a sole proprietorship, partnership, or family corporation."[1] By this definition, the term "family farm" is so ubiquitous as to be meaningless; in 2020, "family farms" accounted for nearly 98 percent of farms in the US. "Small family farms" make up 89 percent of all US farms.[2]

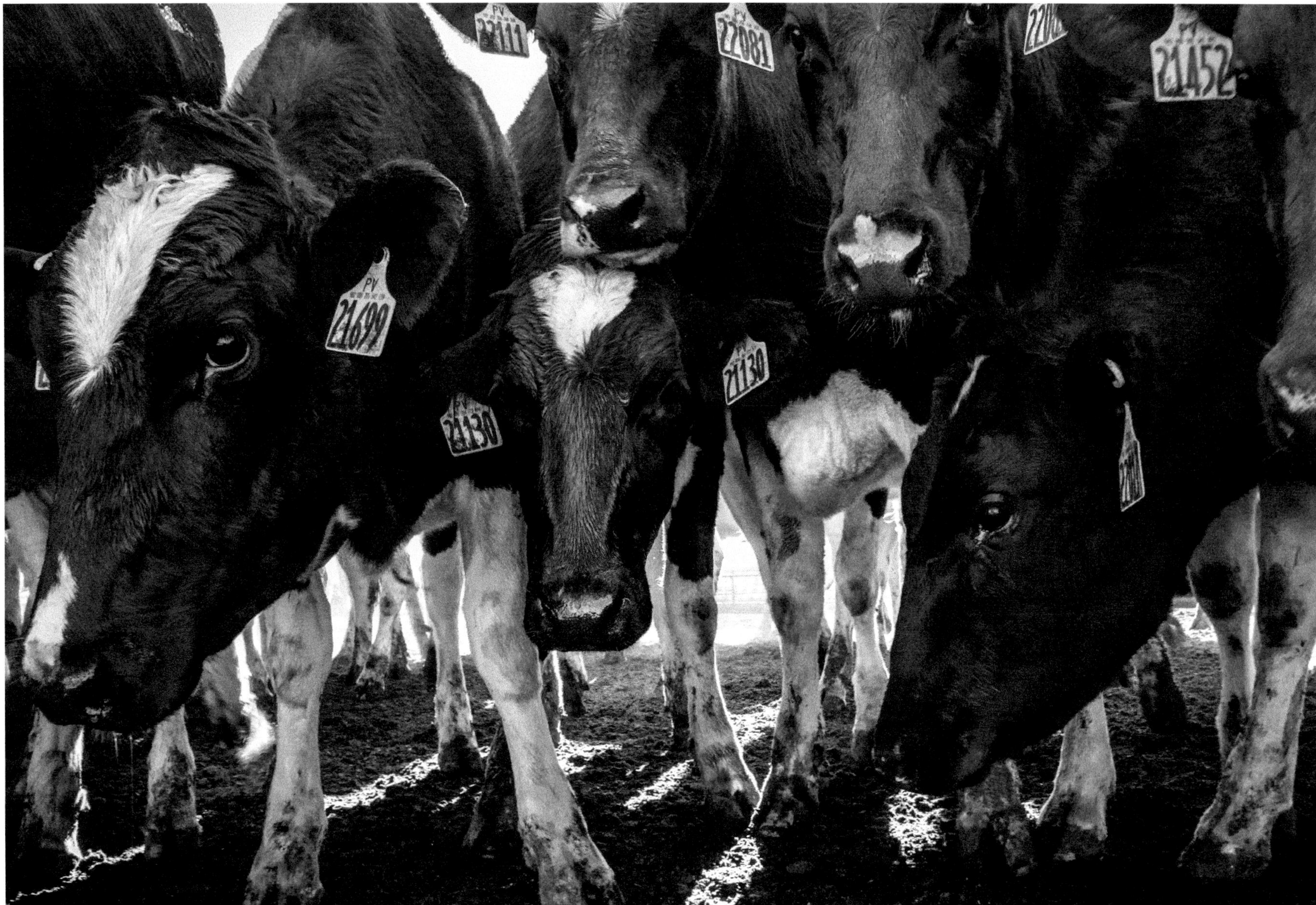

Look Me in the Eyes, 2014

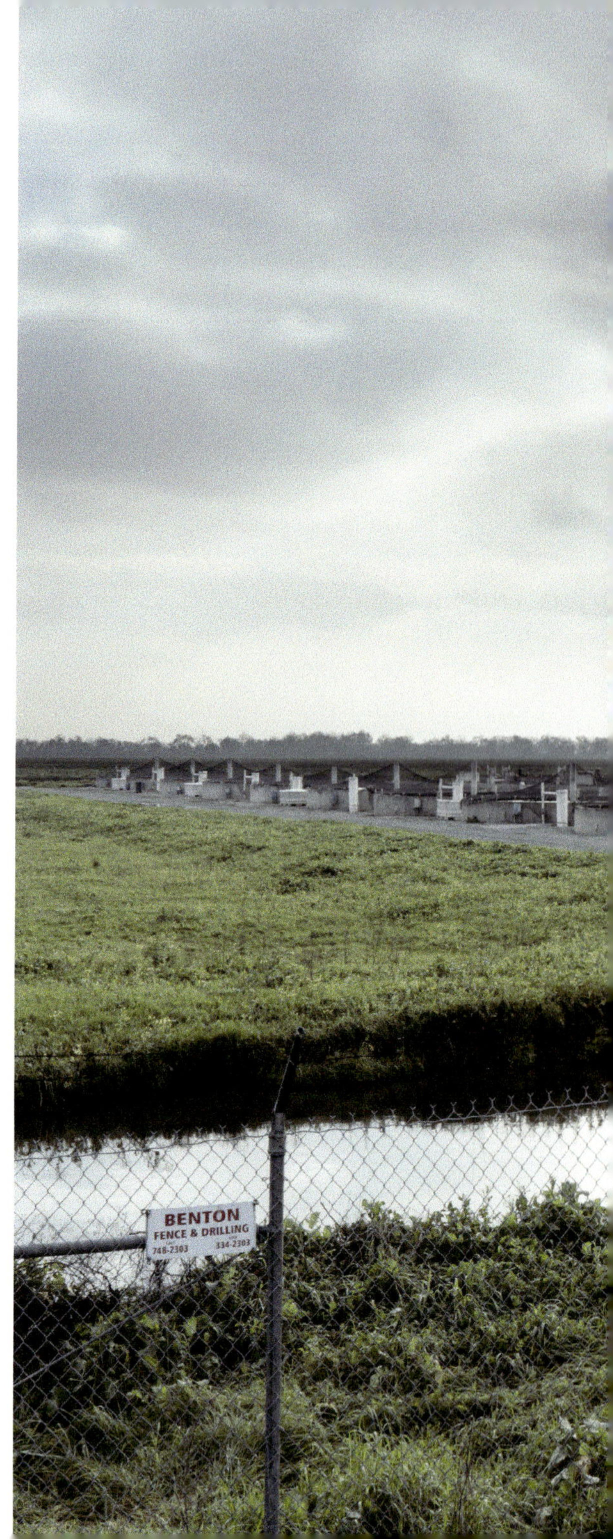

Fish Farm (Sturgeon), 2017
12,000

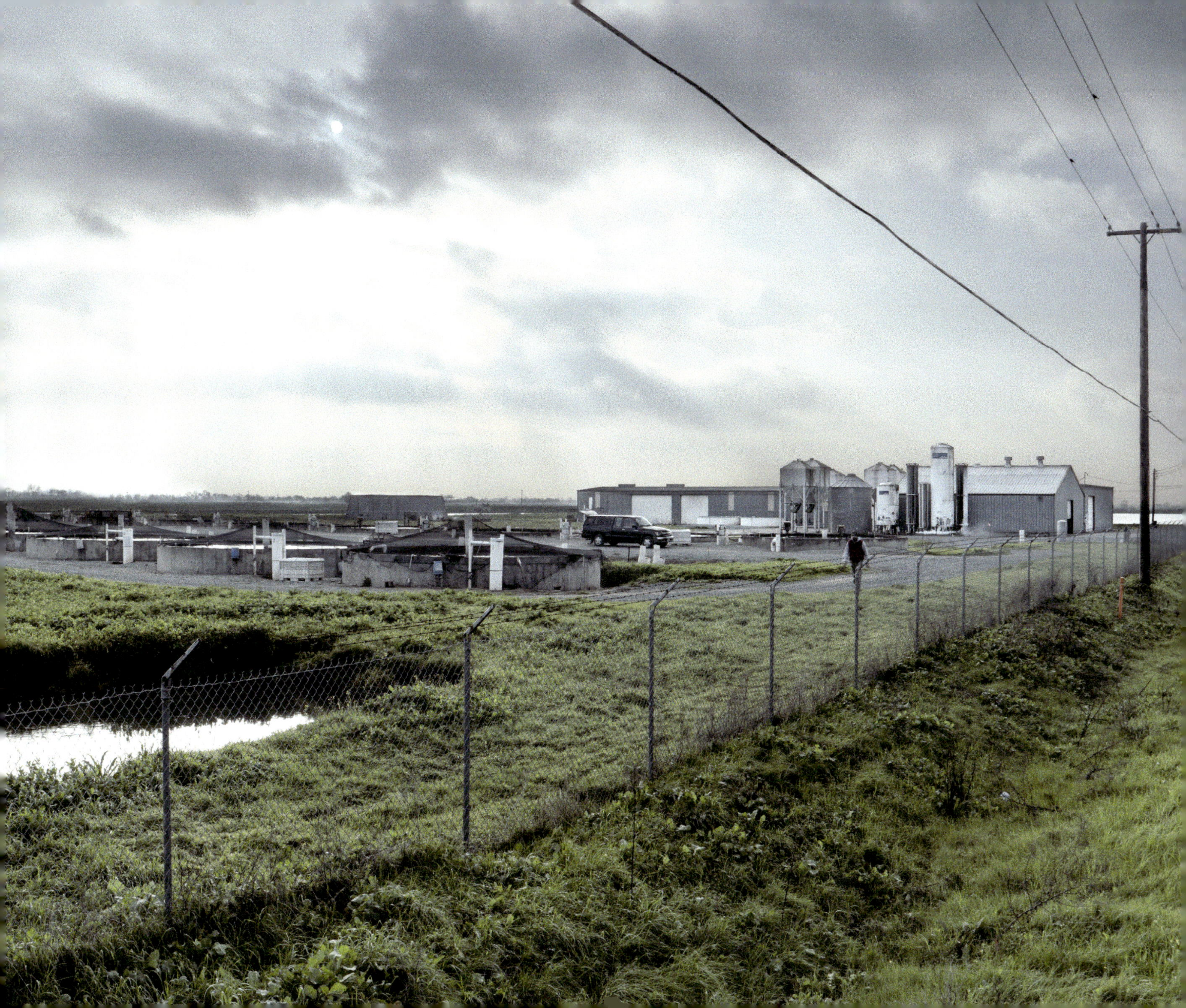

The Lamb

William Blake (1757–1827)

Little Lamb who made thee
    Dost thou know who made thee
Gave thee life & bid thee feed.
By the stream & o'er the mead;
Gave thee clothing of delight,
Softest clothing wooly bright;
Gave thee such a tender voice,
Making all the vales rejoice!
    Little Lamb who made thee
    Dost thou know who made thee

    Little Lamb I'll tell thee,
    Little Lamb I'll tell thee!
He is called by thy name,
For he calls himself a Lamb:
He is meek & he is mild,
He became a little child:
I a child & thou a lamb,
We are called by his name.
    Little Lamb God bless thee.
    Little Lamb God bless thee.[1]

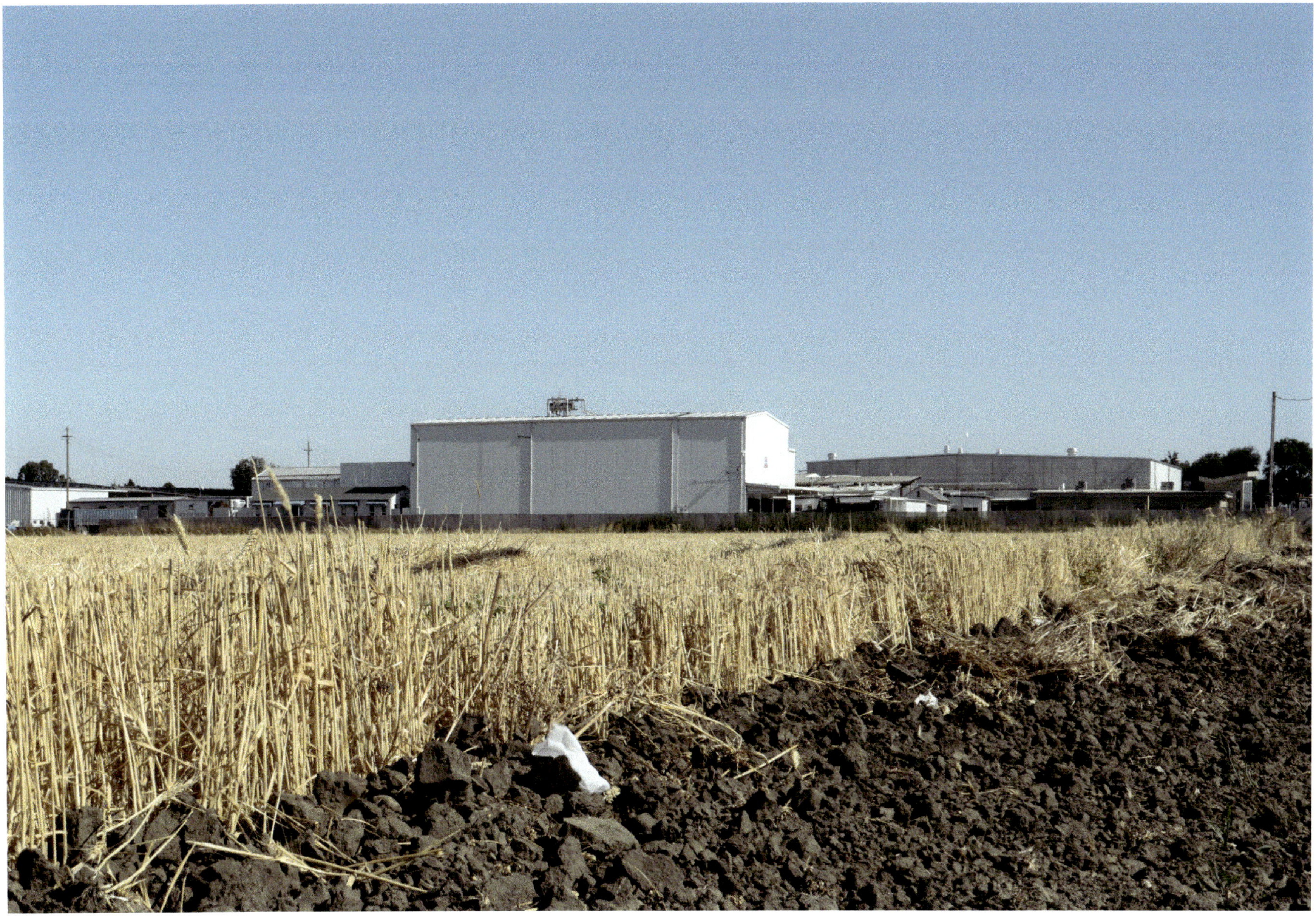

Slaughterhouse (Lambs), 2017
250,000/year

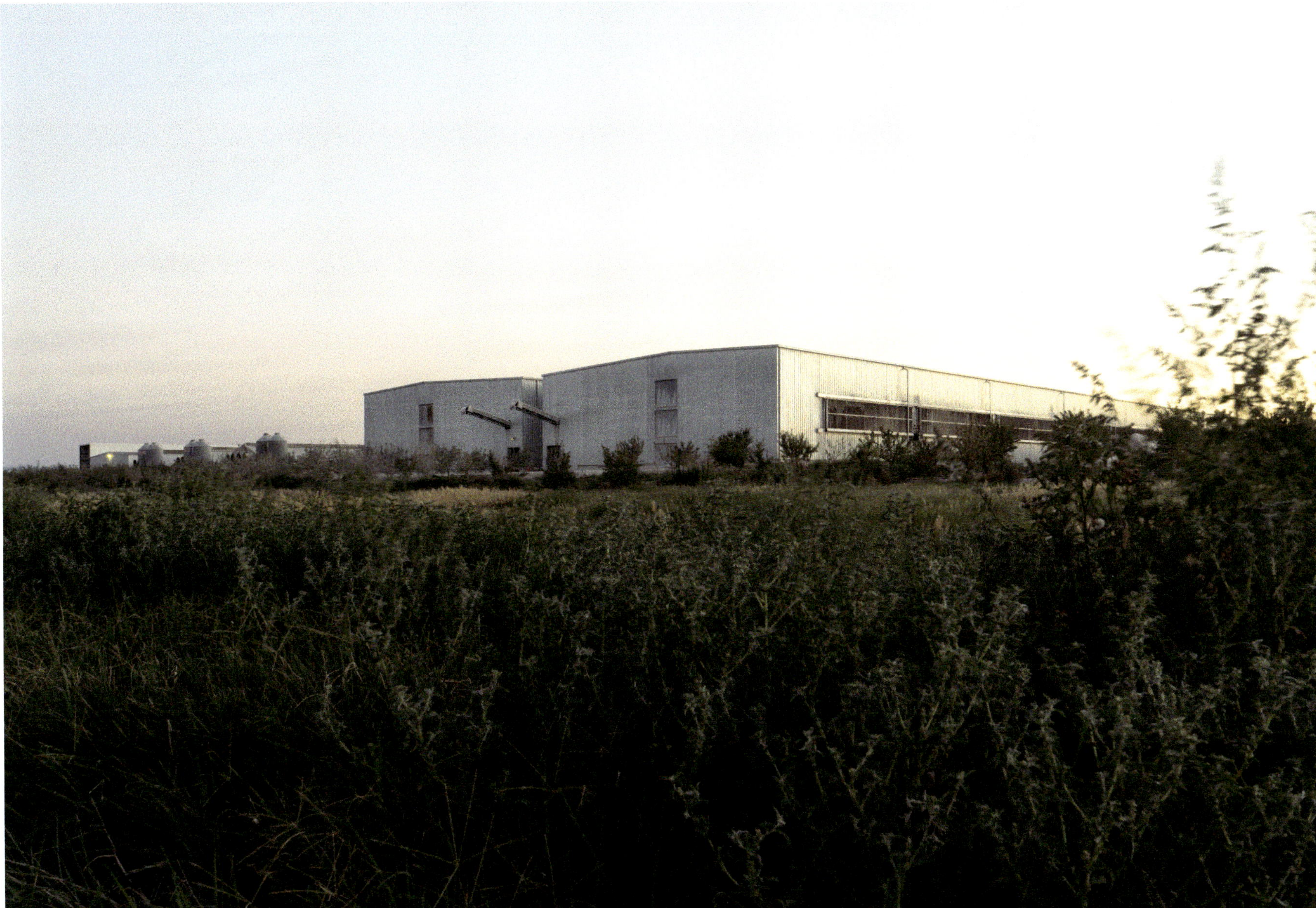

Egg Farm (Chickens), 2017
600,000

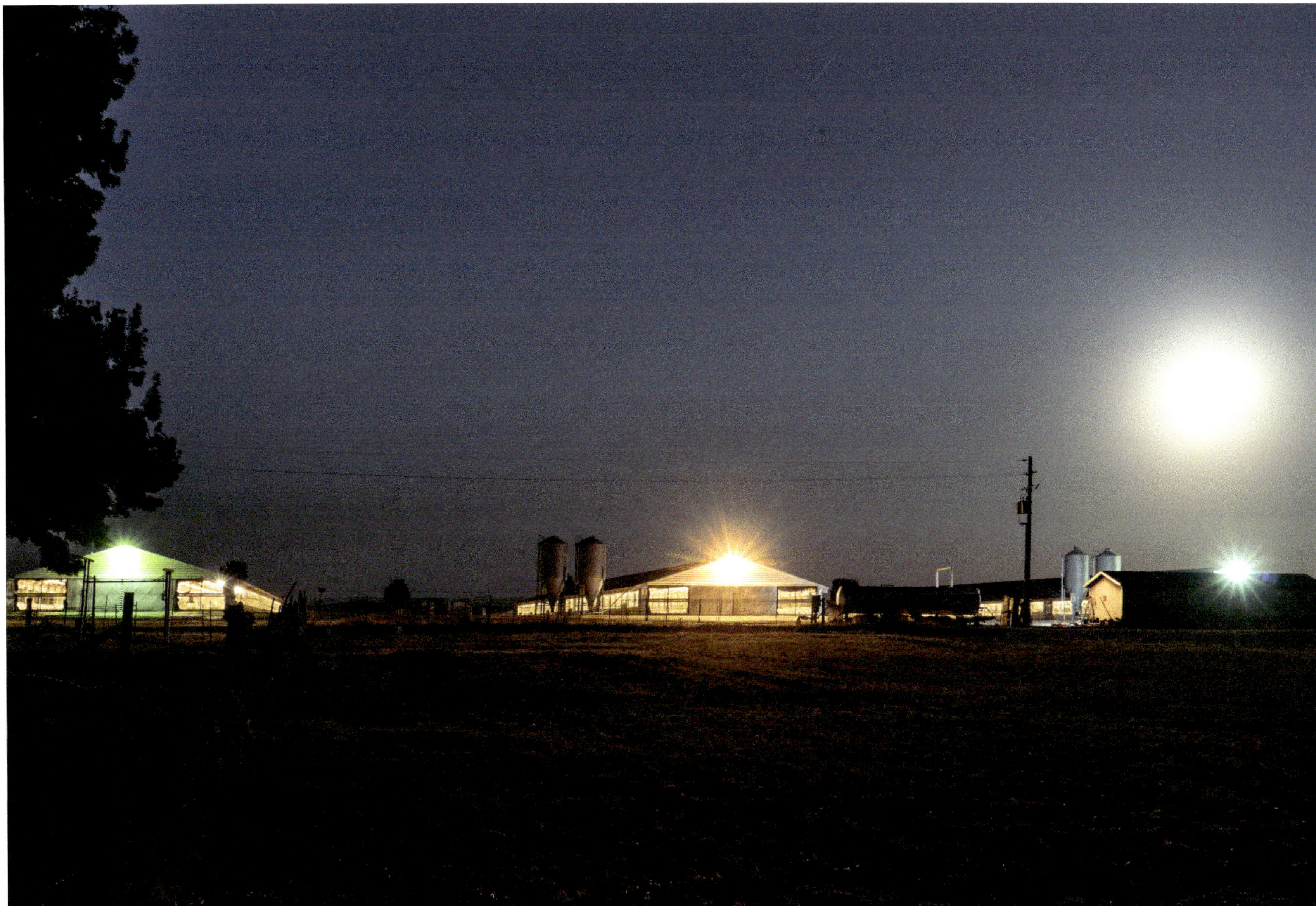

Night of the Full Moon Turkey Farm, 2017
57,000

Sanctuary

The high walls around my grandmother's backyard in Ciudad Juárez, Mexico, were topped with broken glass to keep out the *ladrones*. Dogs and lots of cousins played on the grass and the swing set and the terracotta patio. One door opened to the patio from the kitchen and another from the hall that led to the bedrooms. I spent weeks at a time at my grandmother's house when I was little. When adolescence hit, my immigrant parents didn't know how to deal with their strange American daughter who listened to terrible music, had a bad attitude, and didn't want to go to church anymore. They sent me to spend summers at my grandmother's house.

I remember sitting with my grandmother on the edge of her bed on a Sunday morning—deep blue chenille bedspread, the faint scent of her gentle, powdery perfume. As usual she's wearing a neatly pressed plaid cotton dress, eyeglasses hanging from a thin chain draped around her neck, and an aura of beatific calm. She nestles in my hand a Kennedy half dollar, my *domingo*—Sunday allowance. I kiss her cheek. My grandmother's skin is softer and smoother than the satin coin purse she had given me to stash my *domingos*. I remember my aunt, a benevolent presence who was always doing something—going to work, bringing home groceries, taking out the laundry. She stops everything to greet me, always with an endearment: *mi reina; corazón; mi amor* (my queen; my heart; my love). I remember my cousin teaching me to drive in her little red Toyota Corolla after sunset in an empty parking lot. I'm trying hard not to stall out in first or grind the gears. My cousin screams instructions between raucous, contagious laughter as we lurch around the rows of painted lines. I remember my uncle coming over after work at his ophthalmology practice across the street, sometimes as late as nine or ten. I've already eaten but sit with my uncle as he eats his dinner. Also seated at the table is at least one other cousin. We tell jokes and have meandering conversations about ophthalmology, photography, life, history, and the world.

Those summers at my grandmother's house saved me from 6:30 a.m. alarms, the quagmire that was my parents, and the kids from school with their cliques and mores. The house belonged as much to my grandfather as to my grandmother, but it was always my grandmother's house. The easy orbit of kids and dogs and aunts and uncles originated with my grandmother. The days spent in her home are among my happiest. To this day, I self-soothe by visualizing an architectural walkthrough—the yellow kitchen with a reproduction of *The Last Supper* hanging behind the table; the sitting room with its rows and rows of books and spinning globe; my grandmother's bedroom with its niche for the Virgin Mary statuette. Her home was my sanctuary.

Home, twin sister of sanctuary and refuge, remains within the realm of grandmothers and mothers. Women have always been the homemakers. I'll leave it to others to argue about whether it's innate to our gender or whether it's due to cultural conditioning. What is true is that women have shined in the role. So too, by all accounts the majority of workers at farmed animal sanctuaries are women.[1] Sanctuary staff and volunteers make a home for thousands of the rare survivors of the animal agricultural industry. Making a home, or a sanctuary, involves a tangle of orchestrations, each of which encompasses innumerable jobs: feeding, cleaning, organizing, and on and on. Beyond physical labor, making a home or a sanctuary requires caring about everyone who lives there. Sanctuary workers rehabilitate, care for, love, protect, and support animals in sanctuaries as though they are their own children.

Central as it is to our lives, homemaking doesn't come with the money, recognition, and status of occupations held traditionally by men. "Unpaid work," as it is economically categorized, has been valued at 20 percent of the US Gross Domestic Product,[2] but professional homemakers make less than $13 per hour[3] and over 93 percent are women.[4] Degrees and certificates are scarce when it comes to homemaking. Qualifications include the following: empathetic, nurturing, organized, loving, reliable, affectionate, kind, caring, patient, takes joy in performing endless tiresome tasks because they benefit others—characteristics stereotypically attributed to women. The assumption is that making a home or making a sanctuary should be its own reward, that they don't involve real skills and education, that they come naturally to women. In the animal advocacy movement, the majority of tireless, passionate (and unpaid) activists are women. And yet most well-paid positions of authority are accorded to men.[5] These kinds of inequities aren't limited to animal advocacy; they're pervasive in social justice movements[6] and in society in general.

A subtle and powerful aspect of a patriarchal system is the connection between the domination of women and the exploitation of nature. Ecofeminism exposes the dualism that deprecates "whatever is associated with women—emotions, animals, nature, the body—while elevating those things associated with men—reason, humans, culture, the mind."[7] The sexist hierarchy of "real man" above "natural woman" remains stubbornly ingrained in the human psyche. In 2020, brilliant novelist and essayist Zadie Smith wrote, "(You Make Me Feel Like) A Natural Woman—I used to listen to that song and try to imagine its counterpart. You could make someone feel like a 'real' man—no doubt its own kind of cage—but never a natural one. A man was a man was a man. He bent nature to his will. He did not submit to it."[8] A driving factor in the domination of women, animals, and the earth is a denigrating feminization of nature and naturalizing/animalizing of women.

Through an ecofeminist lens, the correlation between misogyny and animal agriculture comes into sharp focus. Popular metaphors and mean or lewd jokes refer to women as farmed animals: chick, cow, hen, sow. The potential for motherhood means a special kind of torment for farmed animals. A hen in the egg industry is bred to produce exponentially more eggs than is healthy for her body, and she is deprived of light and food to stimulate production (see "Chickens Bred for Their Eggs

FACTS"). A female fish is killed for her eggs, or her eggs are extracted in a painful procedure so she can keep producing caviar for up to ten years.[9] For most of her adult life, a female pig is locked in a metal gestation crate so small she cannot turn around.[10] She will chew on the bars, wave her head incessantly, and manifest other signs of severe distress.[11] While she's lactating, she and her piglets are caged in a metal farrowing crate also too small for her to turn around.[12] Then her piglets are taken from her and she is again inseminated (see "Nonconsensual Sex"). A female calf in the dairy industry spends the first two to three months of her life isolated in a small stall[13] (see *Dairy Farm Calf Hutches, 2016*). As soon as she reaches puberty, she's impregnated; "rape rack" is the colloquial term among dairy workers for the enclosure used to restrain her[14] (see "Nonconsensual Sex"). Soon after giving birth, her calf is taken from her in a heartbreaking ordeal, then she is again impregnated.[15] Mastitis is common in the dairy industry.[16] If a cow, or a pig, or a hen, or a fish could speak a language humans were able to hear, what would she tell us, or more likely, what would she scream?

The imbalance, hubris, tyranny, and violence of a patriarchal culture harms individuals of all genders and species. Male chicks in hatcheries are ground alive or suffocated (see "Chickens Bred for Their Eggs FACTS"). Male calves in the dairy industry are either killed within the first few weeks of life or sold as veal.[17] Almost all species of farmed animals are subjected to artificial insemination, a process that requires masturbating the males and extracting their semen (see "Nonconsensual Sex"). Boys and men, especially those who don't conform to gender norms, are belittled, bullied, or worse. Toxic masculinity robs men and boys of the ability to express a full range of emotions and aspirations. Over 93 percent of US prison inmates are male.[18] A significant factor in the incarceration of men is the influence of hypermasculinity. Prisons are notorious for high rates of violence, sexual assault, and homicide—a recursive loop of victimization.[19]

When I was growing up, steeped in the ideology of the mainstream women's liberation movement, I felt encouraged to pursue careers previously restricted to men. I could be a lawyer! I could be an airline pilot! I could be an astronaut! The sky was the limit! But for the boys around me, so much as a whisper of a desire to pursue "women's work"—homemaker, nurse, nanny—would have been met with smirking derision or would have been crushed entirely. All these years later, change has been slow. In the US, over 71 percent of married women with children under eighteen are in the paid workforce,[20] and yet the bulk of housework is left to women.[21] House husbands are stigmatized and are partners in less than 6 percent of marriages with children.[22]

Ecofeminists propose normalizing an ethic of care—values that revolve around cooperation and compassion, that encourage the kind of efforts necessary in creating homes and sanctuaries out of spaces that would otherwise be oppressive and lonesome, or exploitive and violent.

Imagine flipping the script—subverting the patriarchal, colonizing paradigm. Qualities and occupations trivialized as feminine would be disgendered and generously compensated; sentient nonhuman animals instead of corporations would be granted personhood; forests, rivers, and oceans valued much more than money; prisons and animal farms morphed into homes and sanctuaries. Every transformation, no matter how preposterous or sensible or liberating or urgent, is kindled with sparks of inspiration. Let this be a spark.

May all beings everywhere be joyful and free.

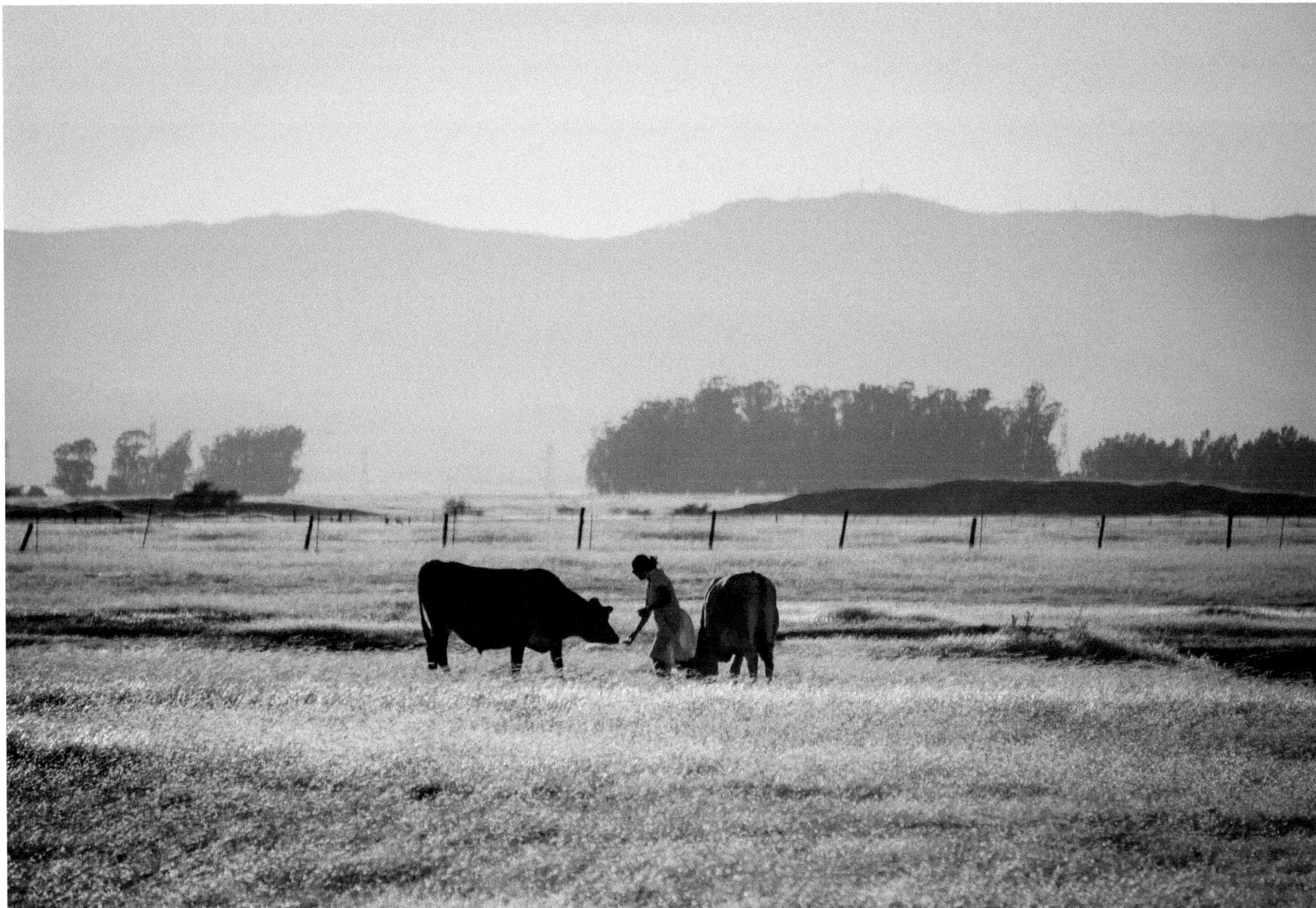

Sanctuary, 2014

Endnotes

Inception

1. "State-By-State Guide to Ag-Gag Legislation," HuschBlackwell, last modified May 22, 2023, https://www.huschblackwell.com/newsandinsights/state-by-state-guide-to-ag-gag-legislation#:~:text=Several%20states%20have%20made%20attempts,Amendment's% 20guarantee%20of%20free%20speech.
2. "Animal Enterprise Terrorism Act," U.S. Government Printing Office, accessed May 26, 2023, https://www.govinfo.gov/content/pkg/PLAW-109publ374/html/PLAW-109publ374.htm.
3. Yinon M. Bar-On, Rob Phillips, and Ron Milo, "The Biomass Distribution on Earth," *Proceedings of the National Academy of Sciences (PNAS)* 115, no. 25 (May 21, 2018): 6506–11, https://doi.org/10.1073/pnas.1711842115.
4. "Livestock Slaughter," United States Department of Agriculture Economics, Statistics and Market Information System, accessed May 26, 2023, https://usda.library.cornell.edu/concern/publications/rx913p88g?locale=en; "Poultry Slaughter," United States Department of Agriculture Economics, Statistics and Market Information System, accessed May 26, 2023, https://usda.library.cornell.edu/concern/publications/3197xm04j?locale=en.

Chickens Bred for Their Eggs FACTS

1. "Red Junglefowl—Meet the Ancestor of the Domesticated Chicken," ChickenMag Poultry & Livestock Magazine, accessed May 31, 2023, https://chickenmag.com/red-junglefowl/.
2. "U.S. Egg Production and Hen Population," United Egg Producers, accessed May 31, 2023, https://unitedegg.com/facts-stats/.
3. C.C. Whitehead, "Skeletal Disorders in Laying Hens: The Problem of Osteoporosis and Bone Fractures," in *Welfare of the Laying Hen*, ed. G.C. Perry (Wallingford Oxfordshire: CABI Publishing, 2004), 259–78.
4. Torgny N. Fredrickson, "Ovarian Tumors of the Hen," *Environmental Health Perspectives* 73 (August 1987): 35–51, https://doi.org/10.1289/ehp.877335.
5. Cheryl Howell, "Egg Binding in Birds—The Symptoms, Causes, and Treatment?" The Happy Chicken Coop, published December 6, 2022, https://www.thehappychickencoop.com/tag/egg-binding-in-birds-treatment/.
6. "A Greener World Technical Advice Fact Sheet No. 13, The Prolific Hybrid Hen: Why Less Is More When It Comes to Egg Laying," A Greener World, published 2009, https://agreenerworld.org/wp-content/uploads/2020/05/TAFS-13-The-Prolific-Hybrid-Hen-v4.pdf.
7. Jonathan Moens, "The Egg Industry Grapples with a Grim Practice: Chick Culling," Undark, published March 15, 2021, https://undark.org/2021/03/15/egg-industry-grapples-with-chick-culling/.
8. Khalil Akhtar, "Male Chick Culling in Egg Industry Could End, Thanks to Canadian Technology," CBC News, last modified June 22, 2016, https://www.cbc.ca/news/science/male-chick-culling-technology-1.3647228.

9. "United Egg Producers Updated Statement on Male Chicks," United Egg Producers, published March 25, 2021, https://unitedegg.com/united-egg-producers-updated-statement-on-male-chicks-2/.
10. "Welfare Implications of Laying Hen Housing," American Veterinary Medical Association, published January 26, 2012, https://www.avma.org/resources-tools/literature-reviews/welfare-implications-laying-hen-housing.
11. "Animal Husbandry Guidelines for U.S. Egg-Laying Flocks," United Egg Producers, published 2017 (page 19), https://uepcertified.com/wp-content/uploads/2021/08/CF-UEP-Guidelines_17-3.pdf; "Cage-free vs. Battery-Cage Eggs," The Humane Society of the United States, accessed July 3, 2023, https://www.humanesociety.org/resources/cage-free-vs-battery-cage-eggs#:~:text=Cage%2DFree%20Systems&text=Unlike%20battery%20hens%2C%20cage%2Dfree,hens%20who%20never%20go%20outside .
12. American Veterinary Medical Association, "Welfare Implications of Laying Hen Housing."
13. United Egg Producers, "U.S. Egg Production and Hen Population."
14. "Questions and Answer—USDA Shell Grading Service," United States Department of Agriculture Agricultural Marketing Service, published October 1, 2015, https://www.ams.usda.gov/publications/qa-shell-eggs.
15. "Cage-Free Verification of USDA Graded Shell Eggs," United States Department of Agriculture, published December 2023, https://www.ams.usda.gov/sites/default/files/AMSFactSheetCageFreeVerification.pdf.
16. American Veterinary Medical Association, "Welfare Implications of Laying Hen Housing."
17. "Does the Label 'Free Range' Pertain to Poultry or Also Other Meats?" United States Department of Agriculture, published March 17, 2023, https://ask.usda.gov/s/article/Does-the-label-free-range-pertain-only-to-poultry-or-also-to-meats.
18. Leigh Weingus, "Are Pasture-Raised Eggs Worth the Splurge? Nutritionists Weigh In," Huffington Post, published August 2, 2021, https://www.huffpost.com/entry/are-pasture-raised-eggs-healthier_l_61005ad3e4b000b997dbbbcc.
19. Christine Nicol, "Feather Pecking and Cannibalism," in *Advances in Poultry Welfare*, ed. Joy A. Mench (Duxford, UK: Woodhead Publishing, 2018), 175–97, https://www.sciencedirect.com/topics/agricultural-and-biological-sciences/debeaking.
20. Sheila E. Scheidler and Sara Shields, "Cannibalism by Poultry," University of Nebraska-Lincoln Extension Publications, published February 2007, https://extensionpublications.unl.edu/assets/html/g1670/build/g1670.htm.
21. Gary D. Butcher and Richard Miles, "Salmonella Control and Molting of Egg-Laying Flocks—Are They Compatible?" Veterinary Medicine—Large Animal Clinical Sciences Department, UF/IFAS Extension, published February 19, 2019, https://edis.ifas.ufl.edu/publication/VM017.
22. "Maximization of Laying Hen Performance, Welfare, Economic Return, and Egg Quality and Safety," United States Department of Agriculture Research, Education & Economics Information System, published September 30, 2013, https://reeis.usda.gov/web/crisprojectpages/0152013-maximization-of-laying-hen-performance-welfare-economic-return-and-egg-quality-and-safety.html.
23. The Editors of Encyclopedia Britannica, "Chicken," *Encyclopedia Britannica*, last modified May 18, 2023, https://www.britannica.com/animal/chicken.
24. "AnAge Entry for *Gallus Gallus*," AnAge: The Animal Ageing and Longevity Database, accessed May 31, 2023, https://genomics.senescence.info/species/entry.php?species=Gallus_gallus.
25. "Humane Methods of Slaughter Act," United States Department of Agriculture National Agricultural Library, accessed July 3, 2023, https://www.nal.usda.gov/animal-health-and-welfare/humane-methods-slaughter-act; "Humane Methods of Slaughter Act," Animal Welfare Institute, accessed July 3, 2023, https://awionline.org/content/humane-methods-slaughter-act.

26. "Animal Welfare Act," United States Department of Agriculture National Agricultural Library, accessed July 3, 2023, https://www.nal.usda.gov/animal-health-and-welfare/animal-welfare-act#:~:text=The%20Animal%20Welfare%20Act%20(AWA,USDA%20%2C%20APHIS%20%2C%20Animal%20Care; David Favre, "Overview of U.S. Animal Welfare Act," Michigan State University College of Law Animal Legal Historical Center, published 2002, https://www.animallaw.info/article/overview-us-animal-welfare-act.

Connection

1. G.W. Lambert, et al., "Effect of Sunlight and Season on Serotonin Turnover in the Brain," *The Lancet* 360, no. 9348 (December 2002): 1840–42, https://doi.org/10.1016/S0140-6736(02)11737-5.
2. Andrew Zimmerman Jones, "What Is the Visible Light Spectrum?" ThoughtCo., last modified February 14, 2020, https://www.thoughtco.com/the-visible-light-spectrum-2699036.
3. Pei-Ju Chen, et al., "Extreme Spectral Richness in the Eye of the Common Bluebottle Butterfly, Graphium Sarpedon," *Frontiers in Ecology and Evolution* 4, no. 18 (March 2016): 1–12, https://doi.org/10.3389/fevo.2016.00018.
4. Yoseph A. Kram, Stephanie Mantey, and Joseph C. Corbo, "Avian Cone Photoreceptors Tile the Retina as Five Independent, Self-Organizing Mosaics," PLOS ONE 5, no. 4 (February 1, 2010), https://doi.org/10.1371/journal.pone.0008992.
5. "Artificial Light Sources in Chicken Houses Increase Egg Laying in Hens During Fall and Winter," Southern Sustainable Agriculture Research and Education, published April 5, 2016, https://southern.sare.org/news/artificial-light-sources-in-chicken-houses-increase-egg-laying-in-hens-during-fall-and-winter/.
6. International Chicken Genome Sequencing Consortium, "Sequence and Comparative Analysis of the Chicken Genome Provide Unique Perspectives on Vertebrate Evolution," *Nature* 432 (2004): 695–716, https://doi.org/10.1038/nature03154; "Researchers Compare Chicken, Human Genomes," National Human Genome Research Institute, published December 8, 2004, https://www.genome.gov/12514316/2004-release-researchers-compare-chicken-human-genomes.
7. Stefano Ghirlanda, Lisolette Jansson, and Magnus Enquist, "Chickens Prefer Beautiful Humans," *Human Nature* 13 (2002): 383–9, https://doi.org/10.1007/s12110-002-1021-6.
8. Lori Marino, "Thinking Chickens: A Review of Cognition, Emotion, and Behavior in the Domestic Chicken," *Animal Cognition* 20, no. 2 (2017): 127–47, https://doi.org/10.1007/s10071-016-1064-4.
9. Marta C. Soares, et al., "Tactile Stimulation Lowers Stress in Fish," *Nature Communications* 2, no. 534 (2011), https://doi.org/10.1038/ncomms1547.
10. C. Sue Carter and Stephen W. Porges, "The Biochemistry of Love: An Oxytocin Hypothesis," *EMBO Reports* 14, no. 1 (2012): 12–16, https://doi.org/10.1038/embor.2012.191.
11. Ed Yong, "Not a Hug Hormone—Fish Version of Oxytocin Acts as Social Spotlight," National Geographic, published October 10, 2012, https://www.nationalgeographic.com/science/article/fish-oxytocin-isotocin-social-spotlight.
12. Chrissy Sexton, "Earth, Humans, and All Living Creatures are Made from Stars," published December 3, 2022, https://www.earth.com/video/earth-humans-and-all-living-creatures-are-made-from-stars/.

Numbers and Names

1. Matilda Holmes, Helena Hamerow, and Richard Thomas, "Close Companions? A Zooarchaeological Study of the Human–Cattle Relationship in Medieval England," *Animals (Basel)* 11, no. 4 (April 20, 2021), https://doi.org/10.3390/ani11041174.
2. "Corporate Personhood," Concise Encyclopedia of Business Ethics, accessed June 8, 2023, https://conciseencyclopedia.org/entries/corporate-personhood/.
3. "Animals to Animal Units Conversion," Iowa Department of Natural Resources, published January 2021, https://www.iowadnr.gov/Portals/idnr/uploads/forms/5420020.pdf.

Chickens Bred for Their Meat FACTS

1. "Poultry Sector at a Glance," United States Department of Agriculture Economic Research Service, accessed July 4, 2023, https://www.ers.usda.gov/topics/animal-products/poultry-eggs/sector-at-a-glance/.
2. E.G. Granquist, et al., "Lameness and Its Relationship with Health and Production Measures in Broiler Chickens," *Animal* 13, no. 10 (2019): 2365–72, https://doi.org/10.1017/S1751731119000466.
3. E.G. Granquist, et al., "Lameness and Its Relationship with Health and Production Measures in Broiler Chickens."
4. Leah Garcés, "Why We Haven't Seen Inside a Broiler Chicken Factory Farm in a Decade," Food Safety News, published January 24, 2013, https://www.foodsafetynews.com/2013/01/why-we-havent-seen-inside-a-broiler-chicken-factory-farm-in-a-decade/.
5. "Poultry Feed Chart and Weight Chart," Agrifarming, last modified August 30, 2023, https://www.agrifarming.in/poultry-feed-chart-and-weight-chart.
6. The Editors of Encyclopaedia Britannica, "Chicken." *Encyclopaedia Britannica*, last modified June 21, 2023, https://www.britannica.com/animal/chicken.
7. "Poultry—Production and Value 2021 Summary," USDA National Agricultural Statistics Service, published April 2022, https://www.uspoultry.org/economic-data/docs/broiler-production-and-value-2021.pdf.
8. "Humane Methods of Slaughter Act," United States Department of Agriculture National Agricultural Library, accessed July 4, 2023, https://www.nal.usda.gov/animal-health-and-welfare/humane-methods-slaughter-act; "Humane Methods of Slaughter Act," Animal Welfare Institute, accessed July 4, 2023, https://awionline.org/content/humane-methods-slaughter-act.
9. "Animal Welfare Act," United States Department of Agriculture National Agricultural Library, accessed July 4, 2023, https://www.nal.usda.gov/animal-health-and-welfare/animal-welfare-act#:~:text=The%20Animal%20Welfare%20Act%20(AWA,USDA%20% 2C%20APHIS%20%2C%20Animal%20Care; David Favre, "Overview of U.S. Animal Welfare Act," Michigan State University College of Law Animal Legal Historical Center, published 2002, https://www.animallaw.info/article/overview-us-animal-welfare-act.
10. "Constituent Update—February 23, 2018, FSIS Criteria for Consideration of Waiver Requests from Young Chicken Slaughter Establishment to Operate at Line Speeds Up to 175 Birds Per Minute," United States Department of Agriculture Food Safety and Inspection Service, accessed July 4, 2023, https://www.fsis.usda.gov/news-events/news-press-releases/constituent-update-february-23-2018-0.

Foie Gras FACTS

1. Herve Remignon and Pierre Burgues, "Evolution of Oxidative Stress Markers in Livers of Ducks During Force-Feeding," *Scientific Reports* 13, no. 1046 (2023), https://doi.org/10.1038/s41598-022-27271-y.
2. Warren Skippon, "The Animal Health and Welfare Consequences of Foie Gras Production," *The Canadian Veterinary Journal*. 54, no. 4 (2013): 403, https://www.ncbi.nlm.nih.gov/pmc/articles/PMC3595949/.
3. "Welfare Aspects of the Production of Foie Gras in Ducks and Geese," Scientific Committee on Animal Health and Animal Welfare, published December 16, 1998, https://food.ec.europa.eu/system/files/2020-12/sci-com_scah_out17_en.pdf.
4. "Welfare Aspects of the Production of Foie Gras in Ducks and Geese." Scientific Committee on Animal Health and Animal Welfare.
5. Skippon, "The Animal Health and Welfare Consequences of Foie Gras Production."
6. "Welfare Aspects of the Production of Foie Gras in Ducks and Geese." Scientific Committee on Animal Health and Animal Welfare.
7. Skippon, "The Animal Health and Welfare Consequences of Foie Gras Production."
8. The Editors of Encyclopedia Britannica, "Anseriform Ecology," Encyclopaedia Britannica, accessed July 2023, https://www.britannica.com/animal/anseriform/Form-and-function.
9. "Amazing Waterfowl Facts," Ducks Unlimited, accessed July 4, 2023, https://www.ducks.org/conservation/waterfowl-research-science/amazing-waterfowl-facts#:~:text=Most%20waterfowl%20fly%20at%20speeds,during%20an%20eight% 2Dhour%20flight.
10. "Welfare Aspects of the Production of Foie Gras in Ducks and Geese," Scientific Committee on Animal Health and Animal Welfare.
11. Jennifer Gaeng, "Duck Lifespan: How Long Do Ducks Live?" last modified June 27, 2023, https://a-z-animals.com/blog/duck-lifespan-how-long-do-ducks-live/.
12. "Amazing Waterfowl Facts," Ducks Unlimited.
13. "Humane Methods of Slaughter Act," USDA National Agricultural Library, accessed July 4, 2023, https://www.nal.usda.gov/animal-health-and-welfare/humane-methods-slaughter-act;"Humane Methods of Slaughter Act," Animal Welfare Institute, accessed July 4, 2023, https://awionline.org/content/humane-methods-slaughter-act.
14. "Animal Welfare Act," USDA National Agricultural Library, accessed July 4, 2023, https://www.nal.usda.gov/animal-health-and-welfare/animal-welfare-act#:~:text=The% 20Animal%20Welfare%20Act%20(AWA,USDA%20%2C%20APHIS%20%2C%20Animal%20Care; David Favre, "Overview of U.S. Animal Welfare Act," Michigan State University College of Law Animal Legal Historical Center, published 2002, accessed March 29, 2023, https://www.animallaw.info/article/overview-us-animal-welfare-act.

Farmed versus Wild Turkeys FACTS

1. M. Shahbandeh, "Total Number of Turkeys Produced in the United States from 1999 to 2021," Statista, published October 10, 2022, https://www.statista.com/statistics/196092/total-number-of-turkeys-raised-in-the-us-since-1999/.
2. "Poultry Slaughter," United States Department of Agriculture Economics, Statistics and Market Information System, last modified June 26, 2023, https://usda.library.cornell.edu/concern/publications/3197xm04j; "Poultry Slaughter, Total Live Weight, and Average Live Weight by

Type," United States Department of Agriculture National Agricultural Statistics Service, published January 25, 2023, https://downloads.usda.library.cornell.edu/usda-esmis/files/3197xm04j/4j03f9428/ht24xw327/psla0123.pdf (page 2).
3. "Wild Turkey Status Update," National Wild Turkey Federation, published November 18, 2022, https://www.nwtf.org/content-hub/wild-turkey-status-update.
4. A.W. Schorger, *The Wild Turkey: Its History and Domestication* (Norman: University of Oklahoma Press 1966): 42–61, https://d3gxcg0i30gmh1.cloudfront.net/095a-98960757-Wild%20Turkey%20Population%20History%20and%20Overview.pdf?versionId=vOorIBvKXRR1_3HOzqp7bW7Ri8J5gOiA#:~:text=Pre%2DColumbian%20populations% 20of%20wild,purposes%20%5B2%2C6%5D.
5. "Wild Turkey Status Update," National Wild Turkey Federation.
6. "Raising America's Turkeys," National Turkey Federation, accessed July 4, 2023, https://www.eatturkey.org/raising-turkeys/.
7. "Wild Turkey Fact Sheet," Connecticut Department of Environmental Protection Wildlife Division, accessed July 4, 2023, https://portal.ct.gov/-/media/DEEP/wildlife/pdf_files/outreach/fact_sheets/wldturkypdf.pdf.
8. "Everything You Want to Know About Turkey Sex," Modern Farmer, published November 16, 2016, https://modernfarmer.com/2016/11/everything-want-know-turkey-sex/#:~:text=How%20Wild%20Turkeys%20Mate,)%20or%20jennies%20(females).
9. "Raising America's Turkeys," National Turkey Federation.
10. "Wild Turkey Fact Sheet," Connecticut Department of Environmental Protection Wildlife Division.
11. "Everything You Want to Know About Turkey Sex," Modern Farmer.
12. Jacy Reese Anthis, "US Factory Farming Estimates," Sentience Institute, last modified April 11, 2019, https://www.sentienceinstitute.org/us-factory-farming-estimates.
13. C.G. Scanes, G. Brant, and M.E. Ensminger, *Poultry Science* 4th Edition (Upper Saddle River: Pearson Prentice Hall, 2004), 270.
14. "Wild Turkey Habitat Management," New York State Department of Environmental Conservation, accessed June 5, 2023, https://www.dec.ny.gov/animals/7288.html#:~:text=Home%20Range&text=A%20turkey% 20can%20and%20does,annual%20needs%20on%20your%20land.
15. Kari Kirschbaum and Jason McCullough, "*Melagris Gallopavo* Wild Turkey," Animal Diversity Web, published 2001, https://animaldiversity.org/site/accounts/information/Meleagris_gallopavo.html#6ffa8ae020c377c6478eb5af38571f6a.
16. "Can Wild Turkeys Fly? (Height, Speed, Distance + FAQs)," Birdfact, last modified July 13, 2022, https://birdfact.com/articles/can-wild-turkeys-fly.
17. "Raising America's Turkeys," National Turkey Federation.
18. "2018 Summary Report on Antimicrobials Sold or Distributed for Use in Food-Producing Animals," United States Food and Drug Administration Center for Veterinary Medicine, published December 2019, https://www.fda.gov/media/133411/download (page 19).
19. Kirschbaum and McCullough, "Melagris Gallopavo Wild Turkey."
20. "Raising America's Turkeys," National Turkey Federation.
21. Kirschbaum and McCullough, "Melagris Gallopavo Wild Turkey."
22. "Modern Turkey Industry," Penn State Extension, accessed July 4, 2023, https://extension.psu.edu/programs/4-h/opportunities/projects/animal-science/poultry/raising-rearing/viii-other-online-resources/the-chicken/modern-turkey-industry.
23. Rose Kivi, "How Does a Turkey Reproduce," Sciencing, last modified April 19, 2018, https://sciencing.com/a-turkey-reproduce-4709177.html.

24. "Raising America's Turkeys," National Turkey Federation.
25. Dave Anderson, Chris Martin, and Andrew Parrella, "Three Cool Facts About Turkeys," Society for the Protection of New Hampshire Forests, published June 9, 2016, https://forestsociety.org/something-wild/three-cool-facts-about-turkeys#:~:text=A% 20turkey%20hen%20 lays%20one,to%20lay%20the%20entire%20clutch.
26. "Humane Methods of Slaughter Act," USDA National Agricultural Library, accessed July 4, 2023, https://www.nal.usda.gov/animal-health-and-welfare/humane-methods-slaughter-act; "Humane Methods of Slaughter Act," Animal Welfare Institute, accessed July 4, 2023, https://awionline.org/content/humane-methods-slaughter-act.
27. "Animal Welfare Act," USDA National Agricultural Library, accessed July 4, 2023, https://www.nal.usda.gov/animal-health-and-welfare/animal-welfare-act#:~:text=The%20Animal%20Welfare%20Act%20(AWA,USDA%20%2C%20APHIS%20%2C%20Animal%20Care; David Favre, "Overview of U.S. Animal Welfare Act," Michigan State University College of Law Animal Legal Historical Center, published 2002, https://www.animallaw.info/article/overview-us-animal-welfare-act.

Photographing Henry

1. A.D. Coleman, *Looking at Photographs: Animals* (San Francisco: Chronicle Books, 1994), 12.
2. Franziska Knolle, Rita P. Goncalves, and A. Jennifer Morton, "Sheep Recognize Familiar and Unfamiliar Human Faces from Two-Dimensional Images," *Royal Society of Open Science* 4, no. 11 (2017): 171228, https://doi.org/10.1098/rsos.171228.
3. John Berger, *About Looking* (New York: Random House Pantheon Books, 1980), 26.

Bad Boss

1. John M. Burger, "Replicating Milgram—Would People Still Obey Today?" *American Psychologist* 64, no. 1 (2009): 1–2, https://doi.org/10.1037/a0010932.
2. Cari Romm, "Rethinking One of Psychology's Most Infamous Experiments," *The Atlantic*, January 28, 2015, https://www.theatlantic.com/health/archive/2015/01/rethinking-one-of-psychologys-most-infamous-experiments/384913/.
3. Stephanie Simon, "A Killing Floor Chronicle," *Los Angeles Times*, December 8, 2003, https://www.latimes.com/archives/la-xpm-2003-dec-08-na-virgil8-story.html.
4. Virgil Butler, "The Cyberactivist," published August 31, 2003, http://cyberactivist.blogspot.com/2003/08/inside-mind-of-killer.html.
5. United Poultry Concerns, "Virgil Butler—'Inside Tyson's Hell: Why I Got Out of the Chicken Slaughtering Business,'" YouTube video, 58:40, recorded August 2004, https://www.youtube.com/watch?v=2vWqOaP_yX8.
6. Shawn Fremstad, Hye Jin Rho, and Hayley Brown, "Meatpacking Workers Are a Diverse Group Who Need Better Protections," Center for Economic and Policy Research, published April 29, 2020, https://cepr.net/meatpacking-workers-are-a-diverse-group-who-need-better-protections/.

7. Rachel M. Macnair, *Perpetration-Induced Traumatic Stress: The Psychological Consequences of Killing* (Westport: Praeger Publishers, 2002), https://books.google.com/books?hl=en&lr=&id=enEOd8YgR0YC&oi=fnd&pg=PP11&dq=perpetration-induced+traumatic+stress+slaughterhouse&ots=9WkMyYNMPa&sig=l1NE9RjCGwyc7m1kJZPUKFGuFio#v=onepage&q&f=false.
8. Jessica Slade and Emma Alleyne, "The Psychological Impact of Slaughterhouse Employment: A Systematic Literature Review," *Trauma, Violence, & Abuse* 24, no. 2 (2021): 429–40; https://doi.org/10.1177/15248380211030243.
9. Karen Victor and Antoni Barnard, "Slaughtering for a Living: A Hermeneutic Phenomenological Perspective on the Well-Being of Slaughterhouse Employees," *International Journal of Qualitative Studies on Health and Well-being* 11, no. 1 (2016): 30266, https://doi.org/10.3402/qhw.v11.30266.
10. Amy J. Fitzgerald, Linda Kalof, and Thomas Dietz, "Slaughterhouses and Increased Crime Rates: An Empirical Analysis of the Spillover From 'The Jungle' Into the Surrounding Community," *Organization & Environment* 22, no. 2 (2009): 158–84, https://doi.org/10.1177/1086026609338164.
11. Andrew Wasley, Christopher D. Cook, and Natalie Jones, "Two Amputations a Week: The Cost of Working in a US Meat Plant, *The Guardian*, July 5, 2018, https://www.theguardian.com/environment/2018/jul/05/amputations-serious-injuries-us-meat-industry-plant#:~:text=US%20meat%20workers%20are%20already,to%20suffer%20repetitive%20strain%20injuries.
12. Abbasali Kasaeinasab, et al., "Respiratory Disorders Among Workers in Slaughterhouses," *Safety and Health at Work* 8, no. 1 (2017): 84–8, https://doi.org/10.1016/j.shaw.2016.04.002.
13. "Workplace Safety and Health: Additional Data Needed to Address Continued Hazards in the Meat and Poultry Industry," United States Government Accountability Office, published April 25, 2016, https://www.gao.gov/products/gao-16-337.
14. "Hands That Feeds Us," Food Chain Workers Alliance, published June 1, 2012, https://foodchainworkers.org/2012/06/the-hands-that-feed-us/; http://foodchainworkers.org/wp-content/uploads/2012/06/Hands-That-Feed-Us-Report.pdf.
15. "The Meatpacking Industry in Rural America During the COVID-19 Pandemic," United States Department of Agriculture Economic Research Service, last modified May 13, 2021, https://www.ers.usda.gov/covid-19/rural-america/meatpacking-industry/.
16. "Poultry Slaughter," United States Department of Agriculture Economics, Statistics and Market Information System, last updated May 25, 2023, https://downloads.usda.library.cornell.edu/usda-esmis/files/3197xm04j/4j03f9428/ht24xw327/psla0123.pdf (page 2).
17. Eder Baron, "Importance of Feed Withdrawal and the Loading Process to the Poultry Industry," WattPoultry, published June 1, 2020, https://www.wattagnet.com/articles/40432-importance-of-feed-withdrawal-and-the-loading-process-to-the-poultry-industry.
18. Dena Jones, "The Welfare of Birds at Slaughter in the United States," Animal Welfare Institute, published November 2020, https://awionline.org/sites/default/files/publication/digital_download/20TheWelfareBirdsSlaughter.pdf (page 3).
19. P.D. Warriss, E.A. Bevis, and S.N. Brown, "Time Spent by Broiler Chickens in Transit to Processing Plants," *The Veterinary Record* 127, no. 25–26 (1990): 617–19, https://pubmed.ncbi.nlm.nih.gov/2089775/.
20. Chris Harris, "Poultry Welfare During Transport," The Poultry Site, published September 3, 2015, https://www.thepoultrysite.com/articles/poultry-welfare-during-transport.
21. Sophie Kevany, "More than Twenty Million Farm Animals Die on Way to Abattoir in US Every Year," *The Guardian*, June 15, 2022, https://www.theguardian.com/environment/2022/jun/15/more-than-20-million-farm-animals-die-on-way-to-abattoir-in-us-every-year.

22. Nico Pitney, "Scientists Believe the Chickens We Eat Are Being Slaughtered While Conscious," Huffington Post, published October 28, 2016, www.huffpost.com/entry/chickens-slaughtered-conscious_n_580e3d35e4b000d0b157bf98.
23. R. Paul Singh and Joe M Regenstein, "Poultry Processing," *Encyclopaedia Britannica*, last modified August 15, 2018, https://www.britannica.com/technology/poultry-processing.
24. "Constituent Update—FSIS Criteria for Consideration of Waiver Requests from Young Chicken Slaughter Establishment to Operate at Line Speeds Up to 175 Birds Per Minute," United States Department of Agriculture Food Safety and Inspection Service, published February 23, 2018, https://www.fsis.usda.gov/news-events/news-press-releases/constituent-update-february-23-2018-0.
25. Charlotte Berg and Mohan Raj, "A Review of Different Stunning Methods for Poultry—Animal Welfare Aspects," *Animals* 5, no. 4 (November 2015): 1207–19, https://www.researchgate.net/publication/285383557_A_Review_of_Different_Stunning_Methods_for_Poultry-Animal_Welfare_Aspects_Stunning_Methods_for_Poultry.
26. "Chickens, Turkeys Being Boiled Alive as Processing Lines Speed Up," *The Washington Post*, last modified August 28, 2017, https://www.dailynews.com/2013/10/30/chickens-turkeys-being-boiled-alive-as-processing-lines-speed-up/.
27. "Humane Methods of Slaughter Act," USDA National Agricultural Library, accessed July 4, 2023, https://www.nal.usda.gov/animal-health-and-welfare/humane-methods-slaughter-act; "Humane Methods of Slaughter Act," Animal Welfare Institute, accessed July 4, 2023, https://awionline.org/content/humane-methods-slaughter-act.
28. "Animal Welfare Act," USDA National Agricultural Library, accessed July 12, 2023, https://www.nal.usda.gov/animal-health-and-welfare/animal-welfare-act#:~:text=The%20Animal%20Welfare%20Act%20(AWA,USDA%20%2C%20APHIS%20%2C%20Animal%20Care; David Favre, "Overview of U.S. Animal Welfare Act," Michigan State University College of Law Animal Legal Historical Center, published 2002, https://www.animallaw.info/article/overview-us-animal-welfare-act.
29. "U.S. Congress, Code § 80502—Transportation of Animals," Office of the Law Revision Counsel, published July 5, 1994, https://uscode.house.gov/view.xhtml?req=granuleid:USC-prelim-title49-section80502&num=0&edition=prelim#sourcecredit.
30. Michelle Pawlinger and Dena Jones, "Animals in Transport Languish as Twenty-Eight Hour Law Goes Off the Rails," *Lewis & Clark School Animal Law Review* 25, no. 1 (2018), https://law.lclark.edu/law_reviews/animal_law_review/past_issues/volume_25_1.php.
31. Timothy Pachirat, *Every Twelve Seconds* (New Haven: Yale University Press, 2013), 145–46.
32. "Livestock Slaughter Procedures," Encyclopedia Britannica, accessed June 12, 2023, https://www.britannica.com/technology/meat-processing/Livestock-slaughter-procedures.
33. Pachirat, *Every Twelve Seconds.*
34. Encyclopedia Britannica, "Livestock Slaughter Procedures."
35. "Humane Slaughter Update," Animal Welfare Institute, published April 2017, https://awionline.org/sites/default/files/products/FA-HumaneSlaughterReport-2017.pdf (6–8, 14).
36. "Modernization of Swine Slaughter Inspection," Federal Register, published October 1, 2019, https://www.federalregister.gov/documents/2019/10/01/2019-20245/modernization-of-swine-slaughter-inspection; Chuck Jolley, "What's Behind the New Swine Slaughter Inspection System?" Food Safety News, published January 10, 2020, https://www.foodsafetynews.com/2020/01/whats-behind-the-new-swine-slaughter-inspection-system/.

37. Brian Handwerk, "An Evolutionary Timeline of *Homo Sapiens*," *Smithsonian Magazine*, February 2, 2021, https://www.smithsonianmag.com/science-nature/essential-timeline-understanding-evolution-homo-sapiens-180976807/.
38. Kurt W. Alt, Ali Al-Ahmad, and Johan Peter Woelber, "Nutrition and Health in Human Evolution—Past to Present," *Nutrients* 14, no. 17 (August 2022): 3594, https://doi.org/10.3390/nu14173594.
39. John Michael Francis, ed., "Columbian Exchange—Livestock," in *Iberia and the Americas: Culture, Politics, and History: A Multidisciplinary Encyclopedia* (Santa, Barbara, CA: ABC-CLIO, 2006), 303–8.
40. "Per Capita Consumption of Poultry and Livestock, 1965 to Forecast 2022, in Pounds," National Poultry Council, last modified December, 2021, https://www.nationalchickencouncil.org/about-the-industry/statistics/per-capita-consumption-of-poultry-and-livestock-1965-to-estimated-2012-in-pounds/.

Of Cash Cows and Golden Eggs FACTS

1. Jared Hayes, "USDA Livestock Subsidies near $50 Billion, EWG Analysis," Environmental Working Group, published February 28, 2022, https://www.ewg.org/news-insights/news/2022/02/usda-livestock-subsidies-near-50-billion-ewg-analysis-finds.
2. Hayes, "USDA livestock subsidies near $50 billion, EWG analysis."
3. Hayes, "USDA livestock subsidies near $50 billion, EWG analysis."
4. "2020 USDA Farm Spending," Agriculture Fairness Alliance, accessed July 4, 2023, https://agriculturefairnessalliance.org/news/2020-usda-farm-program-spending/#easy-footnote-bottom-2-9183.
5. Hayes, "USDA livestock subsidies near $50 billion, EWG analysis."
6. "2020 USDA Farm Spending," Agriculture Fairness Alliance.
7. Kim Souza, "Tyson Foods Maintains Its Top Ranking in Poultry Production," Talk Business & Politics, published March 20, 2019, https://talkbusiness.net/2019/03/tyson-foods-maintains-its-top-ranking-in-poultry-production/.
8. Claire Calloway, "Tyson, JBS, and Other Packers Pay Millions to Settle Poultry and Pork Price-Fixing Suits," Food & Power, published February 4, 2021, https://www.foodandpower.net/latest/title-of-post-yla6k-lmlgr.
9. Dan Papscun, "Pilgrim's Ex-Execs' Charges Dropped in End to DOJ Chicken Probe (2)," Bloomberg Law, last modified October 17, 2022, https://news.bloomberglaw.com/antitrust/ex-pilgrims-execs-get-antitrust-charges-dropped-in-chicken-case.
10. "TYSON FOODS INC -CL A Compensation by Company," Salary.com, accessed July 4, 2023, https://www1.salary.com/TYSON-FOODS-INC-CL-A-Executive-Salaries.html.
11. "Poultry Farm Worker Salary," ZipRecruiter, accessed June 5, 2023, https://www.ziprecruiter.com/Salaries/Poultry-Farm-Worker-Salary#:~:text=Poultry%20Farm%20Worker%20Salary&text=%2425%2C000%20is%20the%2025th%20percentile,Salaries%20below%20this%20are%20outliers.&text=%2435%2C500%20is%20the%2075th%20percentile.
12. Dan Kaufman, "Is It Time to Break Up Big Ag?" *The New Yorker*, April 17, 2021, https://www.newyorker.com/news/dispatch/is-it-time-to-break-up-big-ag.
13. Dan Kaufman, "Is It Time to Break Up Big Ag?"

14. Mark Astley, "Dairy Farmers of America (DFA) Has Agreed to Pay $158.6M to Settle 2007 Lawsuit That Alleged It Conspired with Dean Foods and Others to Suppress Raw Milk Prices across 14 Southeast US States," Dairy Reporter, last modified January 23, 2013, 2023, https://www.dairyreporter.com/Article/2013/01/23/Dairy-Farmers-of-America-agrees-to-156.8m-price-fixing-settlement.
15. Dan Kaufman, "Is It Time to Break Up Big Ag?"
16. Dan Kaufman, "Is It Time to Break Up Big Ag?"
17. "Tom Vilsack," Ballotpedia, accessed July 4, 2023, https://ballotpedia.org/Tom_Vilsack.
18. "Lobbying," Open Secrets, accessed July 4, 2023, https://www.opensecrets.org/industries/lobbying.php?cycle=2022&ind=A04.
19. Carly Fox, et al., "Milked: Immigrant Dairy Farmworkers in New York State," Workers' Center of Central New York and the Worker Justice Center of New York, published 2017, https://milkedny.files.wordpress.com/2017/05/milked_053017.pdf.
20. Flyn Adcock, David Anderson, and Parr Rosson, "The Economic Impacts of Immigrant Labor on U.S. Dairy Farms," Department of Agricultural Economics, Texas A&M University, published August 2015, https://www.nmpf.org/wp-content/uploads/immigration-survey-090915.pdf.
21. "Livestock Grazing on Public Lands," United States Department of the Interior Bureau of Land Management, accessed July 4, 2023, https://www.blm.gov/programs/natural-resources/rangelands-and-grazing/livestock-grazing.
22. Lisa Held, "Just a Few Companies Control the Meat Industry. Can a New Approach to Monopolies Level the Playing Field?" Civil Eats, published July 14, 2021, https://civileats.com/2021/07/14/just-a-few-companies-control-the-meat-industry-can-a-new-approach-to-monopolies-level-the-playing-field/.
23. Lisa Held, "Just a Few Companies Control the Meat Industry. Can a New Approach to Monopolies Level the Playing Field?"
24. "Equilar Virginia CEO Pay Report," Virginia Business, accessed July 4, 2023, https://www.virginiabusiness.com/pdf/1012ceopaychart.pdf.
25. "Slaughterhouse Butcher Salary," Comparably, accessed June 7, 2023, https://www.comparably.com/salaries/salaries-for-slaughterhouse-butcher.
26. Megan Poinski, "Meat Companies Conspired to Set Wages Low, Lawsuit Claims," Divewire by Industry Dive, published November 23, 2022, https://www.fooddive.com/news/meat-wage-lawsuit-filed/637103/.
27. "Dairy: Lobbying," Open Secrets, accessed July 4, 2023, https://www.opensecrets.org/industries/lobbying.php?cycle=2022&ind=A04.
28. "Meat Processing & Products: Lobbying," Open Secrets, accessed July 4, 2023, https://www.opensecrets.org/industries/lobbying.php?ind=G2300.
29. Angela Stuesse and Nathan T. Dollar, "Who Are America's Meat and Poultry Workers?" published September 24, 2020, https://www.epi.org/blog/meat-and-poultry-worker-demographics/.

Eating Animals and Human Health

1. Emily A. Kolbe, "'Won't You Be My Neighbor?' Living with Concentrated Animal Feeding Operations," *Iowa Law Review* 99, no. 1 (November 15, 2013): 427, https://ilr.law.uiowa.edu/print/volume-99-issue/wont-you-be-my-neighbor-living-with-concentrated-animal-feeding-operations.

2. Emily A. Kolbe, "'Won't You Be My Neighbor?' Living with Concentrated Animal Feeding Operations," 427.
3. Chris Dall, "FDA: Antibiotic Use in Food Animals Continues to Rise," University of Minnesota Center for Infectious Disease Research Policy, published December 22, 2016, https://www.cidrap.umn.edu/antimicrobial-stewardship/fda-antibiotic-use-food-animals-continues-rise.
4. "Antibiotic Resistance Threats in the United States," Centers for Disease Control and Prevention, last modified December 2019, https://www.cdc.gov/drugresistance/biggest-threats.html.
5. "PCBs in Farmed Salmon," Environmental Working Group, published July 31, 2003, https://www.ewg.org/research/pcbs-farmed-salmon.
6. An Pan, et al., "Red Meat Consumption and Risk of Type 2 Diabetes: 3 Cohorts of US Adults and an Updated Meta-Analysis," *The American Journal of Clinical Nutrition* 94, no. 4 (October 2011): 1088–96, https://doi.org/10.3945/ajcn.111.018978; Renata Micha, George Michas, and Darius Mozaffarian, "Unprocessed Red and Processed Meats and Risk of Coronary Artery Disease and Type 2 Diabetes—an Updated Review of the Evidence," *Current Atherosclerosis Reports* 14, no. 6 (September 2012): 515–24, https://doi.org/10.1007/s11883-012-0282-8.
7. "Study Strengthens Links Between Red Meat and Heart Disease," European Society of Cardiology, published April 15, 2021, https://www.escardio.org/The-ESC/Press-Office/Press-releases/study-strengthens-links-between-red-meat-and-heart-disease.
8. Harald zur Hausen," Red Meat Consumption and Cancer: Reasons to Suspect Involvement of Bovine Infectious Factors in Colorectal Cancer," *International Journal of Cancer* 130, no. 11 (June 2012): 2475–83, https://doi.org/10.1002/ijc.27413.
9. Y. Wang and M. A. Beydoun, "Meat Consumption Is Associated with Obesity and Central Obesity Among US Adults," *International Journal of Obesity* 33 (2009): 621–28. https://doi.org/10.1038/ijo.2009.45; "Meat Consumption Contributing to Global Obesity," Science Daily, published August 16, 2023, https://www.sciencedaily.com/releases/2016/08/160801093003.htm; Wenpeng You and Maciej Henneberg, "Meat Consumption Providing a Surplus Energy in Modern Diet Contributes to Obesity Prevalence: An Ecological Analysis," *BMC Nutrition* 2, no. 22 (2016), https://doi.org/10.1186/s40795-016-0063-9.
10. Shaghayegh Khodayari, et al., "Meat Consumption and the Risk of General and Central Obesity: The Shahedieh Study," *BMC Research Notes* 15, no. 339 (2022), https://doi.org/10.1186/s13104-022-06235-5.
11. Neal Barnard, Susan Levin, and Caroline Trapp, "Meat Consumption as a Risk Factor for Type 2 Diabetes," *Nutrients* 6, no. 2 (2014): 897–910, https://doi.org/10.3390/nu6020897; "Eating Meat Linked to Higher Risk of Diabetes," Science Daily, published September 5, 2017, https://www.sciencedaily.com/releases/2017/09/170905134506.htm.
12. Pan Zhuang, et al., "Egg and Cholesterol Consumption and Mortality from Cardiovascular and Different Causes in the United States: A Population-Based Cohort Study," *PLoS Medicine* 18, no. 2 (February 2021), https://doi.org/10.1371/journal.pmed.1003508.
13. Christian Løvold Storhaug, Svein Kjetil Fosse, and Lars T. Fadnes, "Country, Regional, and Global Estimates for Malabsorption in Adults: A Systemic Review and Meta-Analysis," *The Lancet Gastroenterology & Hepatology* 2, no. 10 (2017), https://doi.org/10.1016/S2468-1253(17)30154-1.
14. Vânis Viera Borba, et al., "Bovine Milk Proteins as a Trigger for Autoimmune Diseases: Myth or Reality," *International Journal of Celiac Disease* 8, no. 1 (2020), http://pubs.sciepub.com/ijcd/8/1/3/index.html.
15. Dinesh K. Dhanwal, et al., "Epidemiology of Hip Fracture: Worldwide Geographic Variation," *Indian Journal of Orthopaedics* 45, no. 1 (2011): 15–22, https://doi.org/10.4103/0019-5413.73656.
16. Yan Song, et al., "Whole Milk Intake Is Associated with Prostate Cancer-Specific Mortality Among U.S. Male Physicians," *The Journal of Nutrition* 143, no. 2 (2013): 189–96, https://doi.org/10.3945/jn.112.168484; Dagfinn Aune, et al., "Dairy Products, Calcium, and Prostate

Cancer Risk: A Systematic Review and Meta-Analysis of Cohort Studies," *The American Journal of Clinical Nutrition* 101, no. 1 (2015): 87–117, https://doi.org/10.3945/ajcn.113.067157.

17. Gary E. Fraser, et al., "Dairy, Soy, and Risk of Breast Cancer: Those Confounded Milks," *International Journal of Epidemiology* 49, no. 5, (October 2020): 1526–37, https://doi.org/10.1093/ije/dyaa007; "Dairy Consumption Linked to Prostate, Ovarian, Breast Cancers, Finds New Narrative Review," Physicians Committee for Responsible Medicine, published March 7, 2023, https://www.pcrm.org/news/news-releases/dairy-consumption-linked-prostate-ovarian-breast-cancers-finds-new-narrative#:~:text=The%20Adventist%20Health%20Study%2D2,fat%20milks%20produced%20similar%20results.
18. Christian R. Juhl, et al., "Dairy Intake and Acne Vulgaris: A Systematic Review and Meta-Analysis of 78,529 Children, Adolescents, and Young Adults," *Nutrients* 10, no. 8 (August 9, 2018): 1049, https://doi.org/10.3390/nu10081049.
19. "USDA Coexistence Fact Sheets Corn," United States Department of Agriculture, published February 2015, https://www.usda.gov/sites/default/files/documents/coexistence-corn-factsheet.pdf.
20. "USDA Coexistence Fact Sheets Soybeans," United States Department of Agriculture, published February 2015, https://www.usda.gov/sites/default/files/documents/coexistence-soybeans-factsheet.pdf.
21. "Crop Acreage Data," United States Department of Agriculture Farm Service Agency, last modified August, 2022, https://www.fsa.usda.gov/news-room/efoia/electronic-reading-room/frequently-requested-information/crop-acreage-data/index.
22. Jorge Fernandez-Cornejo, "Pesticide Use in U.S. Agriculture: Twenty-One Selected Crops, 1960–2008," USDA Economic Research Service Economic Information Bulletin Number 124, published May 2014, https://www.ers.usda.gov/webdocs/publications/43854/46734_eib124.pdf.
23. "Pesticides in Our Food System," Foodprint, last modified February 8, 2021, https://foodprint.org/issues/pesticides/.
24. "Food System Impacts on Biodiversity Loss," Chatham House, published February 3, 2021, https://www.chathamhouse.org/2021/02/food-system-impacts-biodiversity-loss/summary.
25. Mariah Madigan and Elisa Karhu, " The Role of Plant-Based Nutrition in Cancer Prevention," *Journal of Unexplored Medical Data* 3, no. 9 (2018), http://dx.doi.org/10.20517/2572-8180.2018.05.
26. Claudia De Natale, et al., "Effects of a Plant-Based High-Carbohydrate/High-Fiber Diet versus High-Monounsaturated Fat/Low-Carbohydrate Diet on Postprandial Lipids in Type 2 Diabetic Patients," *Diabetes Care* 32, no. 12 (2009): 2168–73, https://doi.org/10.2337/dc09-0266; Hana Kahleova, et al., "Effect of a Low-Fat Vegan Diet on Body Weight, Insulin Sensitivity, Postprandial Metabolism, and Intramyocellular and Hepatocellular Lipid Levels in Overweight Adults," *JAMA Network Open* 3, no. 11 (2020): e2025454, https://doi.org/10.1001/jamanetworkopen.2020.25454.
27. Gabrielle Turner-McGrievy, Trisha Mandes, and Anthony Crimarco, "A Plant-Based Diet for Overweight and Obesity Prevention and Treatment," *Journal of Geriatric Cardiology* 5, no. 5 (May 2017): 369–74, https://www.ncbi.nlm.nih.gov/pmc/articles/PMC5466943/.
28. Minyang Song, et al., "Association of Animal and Plant Protein Intake With All-Cause and Cause-Specific Mortality," *JAMA Internal Medicine* 176, no. 10 (2016): 1453–63, https://doi.org/10.1001/jamainternmed.2016.4182; Hyunju Kim, et al., "Plant-Based Diets Are Associated With a Lower Risk of Incident Cardiovascular Disease, Cardiovascular Disease Mortality, and All-Cause Mortality in a General Population of Middle-Aged Adults," *Journal of the American Heart Association* 8, no. 16 (2019), https://www.ahajournals.org/doi/10.1161/JAHA.119.012865.

E.T. Phone Home

1. Culum Brown, "Fish Pain: An Inconvenient Truth," *Animal Sentience* 3, no. 32 (2016), https://doi.org/10.51291/2377-7478.1069; Mia Rozenbaum, "Do Fish Feel Pain?" Understanding Animal Research, last modified December 16, 2021, https://www.understandinganimalresearch.org.uk/news/do-fish-feel-pain.
2. "Number of Fish Caught from the Wild Each Year," Fishcount.org.uk, last modified 2019, http://fishcount.org.uk/fish-count-estimates-2/numbers-of-fish-caught-from-the-wild-each-year; "Number of Farmed Fish Slaughtered Each Year," Fishcount.org.uk, last modified 2019, http://fishcount.org.uk/fish-count-estimates-2/numbers-of-farmed-fish-slaughtered-each-year.
3. Chris Sayer, "Troubled Waters|The Women Fighting Sexism in the Fly Fishing Industry," Vice, published August 1, 2019, https://amuse.vice.com/en_us/article/gyaknm/fly-fishing-sexism.
4. "Earthling Project Launches," SETI Institute, published December 17, 2020, https://www.seti.org/press-release/earthling-project-launches.
5. Carlos M. Duarte, et al., "The Soundscape of the Anthropocene Ocean," *Science* 371, no. 6529 (February 5, 2021), https://doi.org/10.1126/science.aba4658.
6. "Marine Plastic Pollution," The International Union for Conservation of Nature, published November 2021, https://www.iucn.org/resources/issues-brief/marine-plastic-pollution#:~:text=Over%20400%20million%20tons%20of,waters%20to%20deep%2Dsea%20sediments.
7. Shubh Mathur, "Dead Zones and Toxic Algae Bloom in U.S. Waters," *Environment & Society Portal, Arcadia*, no. 7 (Spring 2017), https://doi.org/10.5282/rcc/7748).
8. Gideon Mordecai, "Fish Farms Transmit Viruses to Endangered Wild Pacific Salmon, New Evidence Shows," The Conversation, published May 26, 2021, https://theconversation.com/fish-farms-transmit-viruses-to-endangered-wild-pacific-salmon-new-evidence-shows-158691.

Decolonize/Re-Indigenize

1. Emily Rosenbaum and Samantha Friedman, *The Housing Divide—How Generations of Immigrants Fare in New York's Housing Market* (New York: NYU Press, 2006).
2. Alexander Koch, et al., "Earth System Impacts of the European Arrival and Great Dying in the Americas After 1492," *Quaternary Science Reviews* 207 (2019): 13–36, https://doi.org/10.1016/j.quascirev.2018.12.004.
3. Tracy Heim, "The Indigenous Origins of Regenerative Agriculture," National Farmers Union, published October 12, 2020, https://nfu.org/2020/10/12/the-indigenous-origins-of-regenerative-agriculture/; Ethan Gordon, Federico Davila, and Chris Riedy, "Transforming Landscapes and Mindscapes Through Regenerative Agriculture," *Agriculture and Human Values* 39 (2022): 809–26, https://doi.org/10.1007/s10460-021-10276-0.
4. John Michael Francis, ed. "Columbian Exchange—Livestock," in *Iberia and the Americas: Culture, Politics, and History: A Multidisciplinary Encyclopedia* (Santa, Barbara: ABC-CLIO, 2006), 303–8.
5. John W. Frank, Ronald S. Moore, and Genevieve M. Ames, "Historical and Cultural Roots of Drinking Problems Among American Indians," *American Journal of Public Health* 90, no. 3 (2000): 344–51, https://doi.org/10.2105/ajph.90.3.344.

6. "AD 1493–1550s: Native Peoples Begin Dying from European Diseases," National Institutes of Health National Library of Medicine, accessed July 4, 2023, https://www.nlm.nih.gov/nativevoices/timeline/169.html.
7. "Do You Know How Much Water Was Used to Grow Your Food and to Produce Your Clothes and the Things You Buy?" Water Footprint Network, accessed July 4, 2023, https://www.waterfootprint.org/time-for-action/what-can-consumers-do/.
8. Troy Farah, "U.S. Rivers and Lakes Are Shrinking for a Surprising Reason: Cows," *The Guardian*, July 2, 2020, https://www.theguardian.com/environment/2020/jul/02/agriculture-cattle-us-water-shortages-colorado-river.
9. John Derek Scasta, John R. Weir, and Michael C. Stambaugh, "Droughts and Wildfires in Western U.S. Rangelands," *Rangelands* 38, no. 4 (2016): 197–203, https://doi.org/10.1016/j.rala.2016.06.003.
10. Rongbin Xu, et al., "Wildfires, Global Climate Change, and Human Health," *The New England Journal of Medicine* 383 (November 26, 2020): 2173–81, https://doi.org/10.1056/NEJMsr2028985.
11. Casey Kuhn, "'Fire Is Medicine': How Indigenous Practices Could Help Curb Wildfires," Public Broadcasting Service News Hour, published July 23, 2021, https://www.pbs.org/newshour/science/fire-is-medicine-how-indigenous-practices-could-help-curb-wildfires.
12. Heim, "The Indigenous Origins of Regenerative Agriculture."
13. Sunmin Park, Nobuko Hongu, and James W. Daily III, "Native American Foods: History, Culture, and Influence on Modern Diets," *Journal of Ethnic Foods* 3, no. 3 (September 2016): 171–7, https://doi.org/10.1016/j.jef.2016.08.001.
14. Avexnim Cojti, Cesar Gomez, and Bia'ni Madsa' Juárez López, "United in Tradition as Peoples of the Corn," Cultural Survival, published September 4, 2019, https://www.culturalsurvival.org/publications/cultural-survival-quarterly/united-tradition-peoples-corn.
15. Heim, "The Indigenous Origins of Regenerative Agriculture."
16. Jane Mt. Pleasant, "Food Yields and Nutrient Analyses of the Three Sisters: A Haudenosaunee Cropping System," *Ethnobiology Letters* 7, no. 1 (2016): 87–98, https://doi.org/10.14237/ebl.7.1.2016.721.
17. Govindarajan Rajaram, Donald C. Erbach, and Dennis Michael Warren, "The Role of Indigenous Tillage Systems in Sustainable Food Production," *Agriculture and Human Values* 8 (1991): 149–55, https://doi.org/10.1007/BF01579667.
18. Heim, "The Indigenous Origins of Regenerative Agriculture."

Photographing Feedlots

1. Jacy Reese Anthis, "US Factory Farming Estimates," Sentience Institute, last modified April 11, 2019, https://www.sentienceinstitute.org/us-factory-farming-estimates.
2. "Cattle & Beef: Sector at a Glance," United States Department of Agriculture Economic Research Service, last modified September 26, 2022, https://www.ers.usda.gov/topics/animal-products/cattle-beef/sector-at-a-glance/.
3. "Diet and Disease in Cattle: High-Grain Feed May Promote Illness and Harmful Bacteria," American Association for the Advancement of Science—Science Daily, published May 11, 2001, https://www.sciencedaily.com/releases/2001/05/010511074623.htm.

4. "Inhumane Practices on Factory Farms," Animal Welfare Institute, published 2023, https://awionline.org/content/inhumane-practices-factory-farms; "Beef Quality Assurance Supplemental Guidelines," Beef Quality Assurance, published 2014, https://www.bqa.org/Media/BQA/Docs/supplemental_guidelines_2014.pdf.
5. Isaac Basevich Singer, "The Letter Writer," *The New Yorker*, January 13, 1968, 44, https://archives.newyorker.com/newyorker/1968-01-13/flipbook/026/.
6. Josefin Dolstein, "Holocaust Survivor Likens Treatment of Livestock to Shoah," *The Times of Israel*, October 7, 2016, https://www.timesofisrael.com/holocaust-survivor-likens-treatment-of-livestock-to-shoah/.

Excerpt from **"The Eighth Duino Elegy"**

1. Rainer Maria Rilke, *The Poetry of Rainer Maria Rilke* (United Kingdom: CreateSpace Independent Publishing Platform, 2015), 38.

Environmental Impacts

1. "Daily CO_2," CO2-Earth Numbers for Living on Earth, accessed April 23, 2023, https://www.co2.earth/daily-co2#:~:text=420.66%20ppm&text=This%20table%20presents%20the%20most,%3D%20Mauna%20Loa%20Observatory%2C%20Hawaii.
2. Nicola Jones, "How the World Passed a Carbon Threshold and Why It Matters," YaleEnvironment 360, published January 26, 2017, https://e360.yale.edu/features/how-the-world-passed-a-carbon-threshold-400ppm-and-why-it-matters.
3. Melissa Denchak, "Greenhouse Effect 101," National Resources Defense Council, published June 5, 2023, https://www.nrdc.org/stories/greenhouse-effect-101#intro.
4. Sarah Zielinski, "Climate Change Will Accelerate Earth's Sixth Mass Extinction," *Smithsonian Magazine*, April 30, 2015, https://www.smithsonianmag.com/science-nature/climate-change-will-accelerate-earths-sixth-mass-extinction-180955138/.
5. Tammana Begum, "What Is Mass Extinction and Are We Facing a Sixth One?" Natural History Museum, last modified February 21, 2023, https://www.nhm.ac.uk/discover/what-is-mass-extinction-and-are-we-facing-a-sixth-one.html#:~:text=A%20mass%20extinction%20event%20is,less%20than%202.8%20million%20years.
6. Robert Goodland, and Jeff Anhang, "Livestock and Climate Change: What If the Key Actors in Climate Change Are Cows, Pigs, and Chickens?" *World Watch* (November/December 2009): 10–19, https://awellfedworld.org/wp-content/uploads/Livestock-Climate-Change-Anhang-Goodland.pdf.
7. Henning Steinfeld, et al., *Livestock's Long Shadow, Environmental Issues and Options* (Rome: United Nations Food and Agriculture Organization, 2006), https://www.fao.org/3/a0701e/a0701e.pdf.
8. Xiaming Xu, et al., "Global Greenhouse Gas Emissions from Animal-Based Food Are Twice Those of Plant-Based Foods," *Nature Food* 2 (September 2021): 724–32, https://www.nature.com/articles/s43016-021-00358-x.
9. "Methane Emissions Are Driving Climate Change. Here's How to Reduce Them," United Nations Environment Programme, published August 20, 2021, https://www.unep.org/news-and-stories/story/methane-emissions-are-driving-climate-change-heres-how-reduce-them.

10. Joseph Poore and Thomas Nemecek, "Reducing Food's Environmental Impacts through Producers and Consumers," *Science* 360, no. 6392 (2018): 987–992, https://doi.org/10.1126/science.aaq0216.
11. Joseph Poore and Thomas Nemecek, "Reducing Food's Environmental Impacts through Producers and Consumers."
12. Mikaela Weisse and Elizabeth Goldman, "Just Seven Commodities Replaced an Area of Forest Twice the Size of Germany between 2001 and 2015," World Resources Institute, published February 11, 2021, https://www.wri.org/insights/just-7-commodities-replaced-area-forest-twice-size-germany-between-2001-and-2015.
13. Rhett A. Butler, "Consequences of Deforestation," Mongabay, published April 1, 2019, https://rainforests.mongabay.com/09-consequences-of-deforestation.html.
14. "USDA Coexistence Fact Sheets Corn," United States Department of Agriculture, published February 2015, https://www.usda.gov/sites/default/files/documents/coexistence-corn-factsheet.pdf.
15. "USDA Coexistence Fact Sheets Soybeans," United States Department of Agriculture, published February 2015, https://www.usda.gov/sites/default/files/documents/coexistence-soybeans-factsheet.pdf.
16. "Crop Acreage Data," United States Department of Agriculture Farm Service Agency, updated August 2022, https://www.fsa.usda.gov/news-room/efoia/electronic-reading-room/frequently-requested-information/crop-acreage-data/index.
17. "Conserving Plant Genetic Diversity Crucial for Food Security," United Nations News, published October 26, 2010, https://news.un.org/en/story/2010/10/357072#:~:text=FAO%20estimates%20that%2075%20per,because%20of%20a%20changing%20climate.
18. "Biodiversity and Soil Health: How Protecting One, Safeguards the Other," International Union for Conservation of Nature, published May 19, 2022, https://www.iucn.org/news/water/202205/biodiversity-and-soil-health-how-protecting-one-safeguards-other.
19. Alberto Orgiazzi and Panos Panagos, "Soil Biodiversity and Soil Erosion: It Is Time to Get Married," *Global Ecology and Biogeogeography*, 27, no. 10 (2018): 1155–67, https://doi.org/10.1111/geb.12782.
20. Stefano Menegat, Alicia Ledo, and Reyes Tirado, "Greenhouse Gas Emissions from Global Production and Use of Nitrogen Synthetic Fertilisers in Agriculture," *Scientific Reports* 12, no. 14490 (2022), https://doi.org/10.1038/s41598-022-18773-w.
21. Jorge Fernandez-Cornejo, et al., "Pesticide Use in U.S. Agriculture: Twenty-One Selected Crops, 1960–2008," *United States Department of Agriculture Economic Information Bulletin Number 124* (May 2014), https://www.ers.usda.gov/webdocs/publications/43854/46734_eib124.pdf.
22. "Pesticides in Our Food System," Foodprint, last modified February 8, 2021, https://foodprint.org/issues/pesticides/.
23. Tim Benton, et al., "Food System Impacts on Biodiversity Loss: Three Levers for Food System Transformation in Support of Nature," Chatham House, published February 3, 2021, https://www.chathamhouse.org/2021/02/food-system-impacts-biodiversity-loss/03-key-levers-food-system-redesign.
24. Mesfin M. Mekonnen and Arjen Y. Hoekstra, "A Global Assessment of the Water Footprint of Farm Animal Products," *Ecosystems* 15 (2012): 401–15, https://doi.org/10.1007/s10021-011-9517-8.
25. Rutger Willem Hofste, Paul Reig, and Leah Schleifer, "Seventeen Countries, Home to One–Quarter of the World's Population, Face Extremely High Water Stress," World Resources Institute, published August 6, 2019, https://www.wri.org/insights/17-countries-home-one-quarter-worlds-population-face-extremely-high-water-stress.

26. Sandy Milne, "How Water Shortages Are Brewing Wars," BBC Future, published August 16, 2021, https://www.bbc.com/future/article/20210816-how-water-shortages-are-brewing-wars.
27. Jacy Reese Anthis, "US Factory Farming Estimates," Sentience Institute, last modified April 11, 2019, https://www.sentienceinstitute.org/us-factory-farming-estimates.
28. "Over Fifty Groups Petition to Improve Oversight of Water Pollution for Concentrated Animal Feeding Operations," Earthjustice, published October 26, 2022, https://earthjustice.org/press/2022/over-fifty-groups-petition-epa-to-improve-oversight-of-water-pollution-from-concentrated-animal-feeding.
29. Jay P. Graham and Keeve E. Nachman, "Managing Waste from Confined Animal Feeding Operations in the United States: The Need for Sanitary Reform," *Journal of Water Health* 8, no. 4 (2010): 646–670, https://doi.org/10.2166/wh.2010.075;
30. JoAnn Burkholder, et al., "Impacts of Waste from Concentrated Animal Feeding Operations on Water Quality," *Environmental Health Perspectives* 115, no. 2 (2007): 308–12, https://doi.org/10.1289/ehp.8839.
31. Jay P. Graham and Keeve E. Nachman, "Managing Waste from Confined Animal Feeding Operations in the United States: The Need for Sanitary Reform."
32. JoAnn Burkholder, et al., "Impacts of Waste from Concentrated Animal Feeding Operations on Water Quality."
33. Oliver Milman, "Meat Industry Blamed for Largest Ever 'Dead Zone' in Gulf of Mexico," *The Guardian*, August 1, 2017, https://www.theguardian.com/environment/2017/aug/01/meat-industry-dead-zone-gulf-of-mexico-environment-pollution.
34. "What Happens to Animal Waste?" FoodPrint, last modified May 8, 2023, https://foodprint.org/issues/what-happens-to-animal-waste/#easy-footnote-bottom-1-1324.
35. Lucy Ngatia, et al., "Nitrogen and Phosphorous Eutrophication in Marine Ecosystems," in *Monitoring of Marine Pollution*, ed. Houma Bachari Fouzia (London: IntechOpen Limited, 2019), https://doi.org/10.5772/intechopen.81869.
36. Javier Atalah and Pablo Sanchez-Jerez, "Global Assessment of Ecological Risks Associated with Farmed Fish Escapes," *Global Ecology and Conservation* 21 (March 2020), https://doi.org/10.1016/j.gecco.2019.e00842.
37. Ben Panko, "Thousands of Invasive Salmon Escape from Farm in Pacific Northwest," *Smithsonian Magazine*, August 24, 2017, https://www.smithsonianmag.com/smart-news/thousands-invasive-salmon-escape-farm-pacific-northwest-180964619/.
38. Ronald A. Hites, et al., "Global Assessment of Organic Contaminants in Farmed Salmon," *Science* 303, no. 5655 (2004): 226–9, https://doi.org/10.1126/science.1091447.
39. Douglas Frantz and Catherine Collins, "Three Reasons to Avoid Farmed Salmon," *Time*, July 21, 2022, https://time.com/6199237/is-farmed-salmon-healthy-sustainable/.
40. "Protecting Livestock from Predators," United States Department of Agriculture Animal and Plant Health Inspection Service, last modified November 1, 2022, https://www.aphis.usda.gov/aphis/ourfocus/wildlifedamage/operational-activities/SA_Livestock/CT_Protecting_Livestock_Predators.
41. Oliver Millman, "'A Barbaric Federal Program': U.S. Killed 1.75 Million Animals Last Year—Or 200 per Hour," *The Guardian*, March 25, 2022, https://www.theguardian.com/world/2022/mar/25/us-government-wildlife-services-animals-deaths; "Program Data Report G—2021 Animals

Dispersed/Killed or Euthanized/ # Burrows/Dens Removed or Destroyed/ Free or Relocated," United States Department of Agriculture Animal and Plant Health Inspection Service, accessed June 19, 2023, https://www.aphis.usda.gov/aphis/ourfocus/wildlifedamage/SA_Reports/SA_PDRs.
42. Christopher Ketcham, "The Rogue Agency—A USDA Program That Tortures and Kills Endangered Species," *Harpers*, March 2016, https://harpers.org/archive/2016/03/the-rogue-agency/.
43. Michael B. Eisen and Patrick O. Brown, "Rapid Global Phaseout of Animal Agriculture Has the Potential to Stabilize Greenhouse Gas Levels for Thirty Years and Offset 68% Of CO2 Emissions This Century," *Public Library of Science Climate* 1, no. 2 (February 1, 2022), https://doi.org/10.1371/journal.pclm.0000010.
44. Joshua Gibbs and Franceso P. Cappuccio, "Plant-Based Dietary Patterns for Human and Planetary Health," *Nutrients* 14, no. 8 (2022): 1614, https://doi.org/10.3390/nu14081614; Alan Espinosa-Marrón, et al., "Environmental Impact of Animal-Based Food Production and the Feasibility of a Shift Toward Sustainable Plant-Based Diets in the United States," *Frontiers in Sustainability* 3, no. 841106 (May 22), https://doi.org/10.3389/frsus.2022.841106; Alison Pye, et al., "Good for the Planet and Good for Our Health: the Evidence for Whole-Food Plant-Based Diets," *BJPsych International* 19, no. 4 (2022): 90–2, https://doi.org/10.1192/bji.2022.7; Peter Scarborough, et al., "Vegans, Vegetarians, Fish-Eaters and Meat-Eaters in the UK Show Discrepant Environmental Impacts," *NatureFood* (2023), https://doi.org/10.1038/s43016-023-00795-w.

Excerpt from ***The Jungle***

1. Upton Sinclair, *The Jungle* (Champaign, IL: Project Gutenberg, 1906, e-book release date 1994), chap. III, https://www.gutenberg.org/files/140/140-h/140-h.htm.

Grass-Fed Greenwash

1. Allan Savory, "How to Fight Desertification and Reverse Climate Change," TED2013, video 22:19, https://www.ted.com/talks/allan_savory_how_to_fight_desertification_and_reverse_climate_change/transcript?language=en.
2. Julia Calderone, "You May Want to Think Twice Before Buying Expensive Grass-Fed Beef," Business Insider, published March 25, 2016, https://www.businessinsider.com/grass-fed-claims-beef-bogus-usda-packaging-2016-2.
3. James S. Drouillard, "Current Situation and Future Trends for Beef Production in the United States of America—A Review," *Asian-Australasian Journal of Animal Sciences* 31, no. 7 (July 2018): 1007–16, https://doi.org/10.5713/ajas.18.0428.
4. "Grass-fed Beef Production," Penn State Extension, last modified March 7, 2018, https://extension.psu.edu/grass-fed-beef-production.
5. Gene Sager, "Where's Your Beef From? Grass-fed Beef: Is It Green, Humane, & Healthful?" Natural Life Magazine, accessed May 2, 2023, https://www.life.ca/naturallife/0812/grass-fed_beef_green_humane_healthful.htm.
6. "Grass-fed Beef Production," Penn State Extension.
7. Matthew N. Hayek and Rachael D Garrett, "Nationwide Shift to Grass-Fed Beef Requires Larger Cattle Population," *Environmental Research Letters* 13, no. 8 (July 25, 2018), https://doi.org/10.1088/1748-9326/aad401.
8. "Grass Fed vs. Grain Fed Beef," Penn State Extension, accessed May 3, 2023, https://extension.psu.edu/grass-fed-vs-grain-fed-beef.

9. Christopher D. Lupo, et al., "Life-Cycle Assessment of the Beef Cattle Production System for the Northern Great Plains, USA," *Journal of Environmental Quality* 42, no. 5 (September–October 2013), https://doi.org/10.2134/jeq2013.03.0101.
10. Erin Jones, Ariane Datil, and Alanna Delfino, "Yes, Cattle Are the Top Source of Methane Emissions in the U.S.," Verify, last modified December 14, 2021, https://www.verifythis.com/article/news/verify/environment-verify/cattle-cows-the-top-source-of-methane-emissions-in-united-states/536-8d5bf326-6955-4a9c-8ea5-761d73ba464c.
11. "Importance of Methane," United States Environmental Protection Agency Global Methane Initiative, last modified May 22, 2023, https://www.epa.gov/gmi/importance-methane#:~:text=Methane%20is%20more%20than%2025,due%20to%20human%2Drelated%20activities.
12. Ralph Maughn, "Alan Savory Gives a Popular and Very Misleading Ted Talk," The Wildlife News, published March 18, 2013, https://www.thewildlifenews.com/2013/03/18/alan-savory-gives-a-popular-and-very-misleading-ted-talk/; James McWilliams, "All Sizzle and No Steak," Slate, published April 22, 2013, https://slate.com/human-interest/2013/04/allan-savorys-ted-talk-is-wrong-and-the-benefits-of-holistic-grazing-have-been-debunked.html; John Carter, et al., "Holistic Management: Misinformation on the Science of Grazed Ecosystems," *International Journal of Biodiversity* 2014, no. 163431 (April 23, 2014), https://doi.org/10.1155/2014/163431; George Monbiot, "Eat More Meat and Save the World: The Latest Implausible Farming Miracle," *The Guardian*, August 4, 2014, https://www.theguardian.com/environment/georgemonbiot/2014/aug/04/eat-more-meat-and-save-the-world-the-latest-implausible-farming-miracle; Maria Nordborg, *Holistic Management—A Critical Review of Allan Savory's Grazing Method* (Uppsala: SLU/EPOK—Centre for Organic Food & Farming & Chalmers, 2016), https://orgprints.org/id/eprint/34330/1/holisticmanagement_review.pdf; Christopher Ketcham, "Allan Savory's Holistic Management Theory Falls Short on Science," *Sierra Club Magazine*, February 23, 2017, https://www.sierraclub.org/sierra/2017-2-march-april/feature/allan-savory-says-more-cows-land-will-reverse-climate-change; "New Rebuttal to the Myth 'Holistic Management Can Reverse Climate Change,'" ClimateChange.ie, published March 18, 2019, http://www.climatechange.ie/new-rebuttal-to-the-myth-holistic-management-can-reverse-climate-change/.
13. Jared Kukura, "Theoretical Ecological Collapse: Any Excuse to Kill Elephants," *The Journal of African Elephants*, published April 2, 2020, https://www.africanelephantjournal.com/theoretical-ecological-collapse-any-excuse-to-kill-elephants/.
14. "Back to Grass the Market Potential for U.S. Grass-Fed Beef," Stone Barns Center, published April 2017, https://www.stonebarnscenter.org/wp-content/uploads/2017/10/Grassfed_Full_v2.pdf.
15. Christopher L. Weber and H. Scott Matthews, "Food-Miles and the Relative Climate Impacts of Food Choices in the United States," *Environmental Science & Technology* 42, no. 10 (2008): 3508–13, https://doi.org/10.1021/es702969f.
16. John Michael Francis, ed. "Columbian Exchange—Livestock," *Iberia and the Americas: Culture, Politics, and History: A Multidisciplinary Encyclopedia* (Santa Barbara, CA: ABC-CLIO, 2006), 303–8.
17. James M. MacDonald and William D. McBride, "The Transformation of U.S. Livestock Agriculture Scale, Efficiency, and Risks," *Economic Research Service Economic Information Bulletin Number 43* (USDA January 2009), https://www.ers.usda.gov/webdocs/publications/44292/10992_eib43.pdf?v=0.
18. "Why Grass-Fed Beef Costs More," Consumer Reports, published August 24, 2015, https://www.consumerreports.org/cro/magazine/2015/08/why-grass-fed-beef-costs-more/index.htm.

Nonconsensual Sex

1. Stephen J. Dubner, "Artificial Insemination: What About the Other Animals?" Freakonomics, published November 22, 2011, https://freakonomics.com/2011/11/artificial-insemination-what-about-the-other-animals/.
2. Keith Bramwell, "Artificial Insemination in Turkeys and Chickens," Merck Veterinary Manual, last modified October 2022, https://www.merckvetmanual.com/poultry/artificial-insemination/artificial-insemination-in-turkeys-and-chickens#:~:text=In%20most%20commercial%20chicken%20production,breeding%20work%20for%20research%20purposes.
3. Dubner, "Artificial Insemination: What About the Other Animals?"
4. "Economic Impact of Artificial Insemination vs. Natural Mating for Beef Cattle Herds," Publication Number P2486, Mississippi State University Extension, last modified 2020, http://extension.msstate.edu/publications/economic-impact-artificial-insemination-vs-natural-mating-for-beef-cattle-herds.
5. Michael L. O'Connor and Jana L. Peters, "Artificial Insemination Technique: Cattle," Penn State Extension, last modified May 13, 2022, https://extension.psu.edu/artificial-insemination-technique-cattle.
6. A. De Vries and M.I. Marcondes, "Review: Overview of Factors Affecting Productive Lifespan of Dairy Cows," *Animal* 14, no. S1 (2020): s155–s164, https://doi.org/10.1017/S1751731119003264.
7. Melissa Rouge, "Semen Collection from Bulls," Colorado State University, last modified September 2, 2002, http://www.vivo.colostate.edu/hbooks/pathphys/reprod/semeneval/bull.html.
8. B.K. Whitlock, et al., "Electroejaculation Increased Vocalization and Plasma Concentrations of Cortisol and Progesterone, but Not Substance P, in Beef Bulls," *Theriogenology* 78 (2012): 737–46, https://doi.org/10.1016/j.theriogenology.2012.03.020.
9. Dubner, "Artificial Insemination: What About the Other Animals?"
10. Dominiek Maes, et al., "Artificial Insemination in Pigs," in *Artificial Insemination in Farm Animals*, ed. Milad Manafi (London: InTech, 2011), Introduction, https://doi.org/10.5772/16592.
11. Malcolm Rainey and Paul Vining, "Artificial Insemination in Swine: Breeding the Female," published June 13, 2022, https://www.aces.edu/blog/topics/farming/artificial-insemination-in-swine-breeding-the-female/.
12. D. Steven Pollmann, "Sow Lifetime Productivity: Importance of Monitoring in Commercial Pork Production," *Allen D. Leman Swine Conference* 39 (2012), https://conservancy.umn.edu/bitstream/handle/11299/139636/Pollmann.pdf?sequence=1&isAllowed=y.

It's a Family Affair

1. "Family Farms," United States Department of Agriculture National Institute of Food and Agriculture, accessed June 27, 2023, https://www.nifa.usda.gov/grants/programs/family-small-farm-program/family-farms#:~:text=USDA%20classifies%20family%20farms%20as,as%20farms%20with%20hired%20managers%E2%80%9D.
2. "Farm Structure and Contracting," United States Department of Agriculture Economic Research Service, last modified March 8, 2022, https://www.ers.usda.gov/topics/farm-economy/farm-structure-and-organization/farm-structure-and-contracting/.

The Lamb

1. William Blake, "Songs of Innocence–Song 8," in *The Complete Poetry & Prose of William Blake*, ed. David V. Erdman (New York: Anchor Books Doubleday), 8–9.

Sanctuary

1. In my considerable experience at farmed animal sanctuaries from 2013 to 2023, the great majority of staff and volunteers have been women. An informal poll of seven farmed animal sanctuaries in the summer of 2023, including three of the largest in the US, confirmed my observations.
2. "Transforming Economies, Realizing Rights," UN Women—United Nations Entity for Gender Equality and the Empowerment of Women, published 2015, https://progress.unwomen.org/en/2015/pdf/UNW_progressreport.pdf.
3. "Homemaker Salary," Zippia, last modified April 6, 2023, https://www.zippia.com/homemaker-jobs/salary/.
4. "Homemaker Demographics and Statistics in the US," Zippia, published 2021, https://www.zippia.com/homemaker-jobs/demographics/.
5. Harold A. Herzog, "Gender Differences in Human-Animal Interactions: A Review," *Anthrozoos* 20, no. 1 (2007), https://cdn2.psychology-today.com/assets/attachments/44240/gender-and-human-animal-interactions.pdf; Lisa Kemmerer, "Evidence of Sexism and Male Privilege in the Animal Liberation/Rights Movement," *Between the Species*, 21, no. 1 (Spring 2018), http://lisakemmerer.com/articles/evidence_of_sexism_and_male_privilege_in_the_animal_liberation_rights_movement.pdf; Emily Gaarder, *Women and the Animal Rights Movement*, (New Brunswick: Rutgers University Press, 2011).
6. Lisa Kemmerer, "The 'Why' of Sexism in Social Justice Movements," in *Oppressive Liberation* (London: Palgrave Macmillan, 2023), 69–93, https://doi.org/10.1007/978-3-031-15363-1_4.
7. "Quick Facts United States," United States Census Bureau, published 2021, https://www.census.gov/quickfacts/fact/table/US/LFE046221.
8. Emma Hinchliffe, "Women CEOs Run 10.4% of Fortune 500 Companies. A Quarter of the Fifty-Two Leaders Became CEO in the Last Year," *Fortune*, June 5, 2023, https://fortune.com/2023/06/05/fortune-500-companies-2023-women-10-percent/.
9. Rebecca Leppert and Drew Desilver, "118th Congress Has a Record Number of Women," Pew Research Center, published January 3, 2023, https://www.pewresearch.org/short-reads/2023/01/03/118th-congress-has-a-record-number-of-women/#:~:text=Women%20make%20up%20more%20than,stood%20even%20a%20decade%20ago.
10. "Native Women and the Pay Gap," American Association of University Women, published 2021, https://www.aauw.org/resources/article/native-women-and-the-pay-gap/.
11. "Women of Color and the Wage Gap," Center for American Progress, published November 17, 2021, https://www.americanprogress.org/article/women-of-color-and-the-wage-gap/.
12. "National Population by Characteristics: 2020–2022," United States Census Bureau, published 2021, https://www.census.gov/data/datasets/time-series/demo/popest/2020s-national-detail.html.
13. Katherine Schaeffer, "Twenty-Two States Have Ever Elected a Black Woman to Congress," Pew Research Center, published February 16, 2023, https://www.pewresearch.org/short-reads/2023/02/16/22-states-have-ever-elected-a-black-woman-to-congress/#:~:text=Sixty%2Done%20women%20of%20color,and%2011%20are%20Asian%20American.

14. "More Women Are Becoming Fortune 500 CEOs, But Only Three Of Them Are WOC," People of Color in Tech, published January 27, 2023, https://peopleofcolorintech.com/articles/more-women-are-becoming-fortune-500-ceos-but-only-three-of-them-are-woc/.
15. "Gender Inequality Facts & Figures," Social Justice Resource Center, accessed June 23, 2023, https://socialjusticeresourcecenter.org/facts-and-figures/gender-inequality/.
16. Lin Bian, Sarah-Jane Leslie, and Andrei Cimpian, "Evidence of Bias against Girls and Women in Contexts That Emphasize Intellectual Ability," *American Psychologist* 73, no. 9 (2018): 1139–53, https://doi.org/10.1037/amp0000427; Daniel Storage, et al., "Adults and Children Implicitly Associate Brilliance with Men More than Women," *Journal of Experimental Social Psychology* 90, no. 104020 (2020), https://doi.org/10.1016/j.jesp.2020.104020; Amarette Filut, Anna Kaatz, and Molly Carnes, "The Impact of Unconscious Bias on Women's Career Advancement," *The Sasakawa Peace Foundation Expert Reviews Series on Advancing Women's Empowerment* (2007), https://www.spf.org/en/global-image/units/upfiles/163038-1-20211216165119_b61baeff7486a9.pdf; Debra M. Easterly and Cynthia S. Ricard, *Journal of Research Administration* 42, no. 1 (Spring 2011): 61–73, https://files.eric.ed.gov/fulltext/EJ955003.pdf.
17. "Get the Facts—Learn About Gender Inequity in the Arts with Some Eye-Opening Facts," National Museum of Women in the Arts, accessed July 11, 2023, https://nmwa.org/support/advocacy/get-facts/; Julia Halperin and Charlotte Burns, "Female Artists Represent Just 2 Percent of the Market. Here's Why—and How That Can Change," Artnet, published September 19, 2019, https://news.artnet.com/womens-place-in-the-art-world/female-artists-represent-just-2-percent-market-heres-can-change-1654954.
18. Greta Gaard, "Living Interconnections with Animals and Nature," in *Ecofeminism Women, Animals, Nature,* ed. Greta Gaard (Philadelphia: Temple University Press, 1993), 5.
19. Zadie Smith, *Intimations Six Essays* (London: Penguin Books, 2020), 4.
20. David Stephen, "A Tale of Two Caviars—Caviar and No-Kill Caviar," Marshallberg Farm, published May 6, 2022, https://thecaviarfarm.com/caviar-and-no-kill-caviar/.
21. "Gestation Crates: A Growing Financial Risk," Fairr A Coller Initiative, published February 22, 2022, https://www.fairr.org/article/gestation-crates/.
22. Chris Peterson, "Pigs Go Insane In Gestation Crates," The Asbury Park Press, November 26, 2014, https://www.app.com/story/opinion/columnists/2014/11/26/petersen-pigs-go-insane-gestation-crates/19542201/.
23. Emma M. Baxter, Inger Lise Andersen, and Sandra A. Edwards, "2—Sow Welfare in the Farrowing Crate and Alternatives," in *Advances in Pig Welfare,* ed. Marek Špinka (Sawston, UK: Woodhead Publishing, 2018), 27–72, https://doi.org/10.1016/B978-0-08-101012-9.00002-2.
24. Bonnie Beaver and Gail Golab, "Livestock Welfare Issues," in *The Veterinarian's Guide to Animal Welfare,* ed. Bonnie Beaver and Gail Golab (Cambridge, MA: Academic Press, 2023), https://www.sciencedirect.com/topics/agricultural-and-biological-sciences/dairy-calves.
25. James McWilliams, "Milk of Human Kindness," *Forbes,* October 25, 2013, https://www.forbes.com/sites/jamesmcwilliams/2013/10/25/milk-of-human-kindness-denied-to-dairy-cows/?sh=7dadeae6598b.
26. Sarah E. Bolton and Marina A.G. von Keyserlingk, "The Dispensable Surplus Dairy Calf: Is This Issue a 'Wicked Problem' and Where Do We Go from Here?" *Frontiers in Veterinary Science* 8 (April 14, 2021), https://doi.org/10.3389/fvets.2021.660934.
27. Juliano L. Gonçalves, et al., "Incidence and Treatments of Bovine Mastitis and Other Diseases on Thirty-Seven Dairy Farms in Wisconsin," *Pathogens* 11, no. 11 (2022): 1282, https://doi.org/10.3390/pathogens11111282.

28. Bolton and von Keyserlingk, "The Dispensable Surplus Dairy Calf: Is This Issue a 'Wicked Problem' and Where Do We Go from Here?"
29. "Inmate Gender," Federal Bureau of Prisons, last modified June 24, 2023, https://www.bop.gov/about/statistics/statistics_inmate_gender.jsp.
30. Mariah Woodson, "The Best We Can Be: How Toxic Masculinity Creates a Second Inescapable Situation for Inmates," *Public Interest Law Reporter* 24, no. 2 (2019): 2, https://lawecommons.luc.edu/cgi/viewcontent.cgi?article=1569&context=pilr.
31. "Employment Characteristics of Families—2022," Bureau of Labor Statistics, published April 19, 2023, https://www.bls.gov/news.release/pdf/famee.pdf (page 2).
32. Damla Onder, "Women's Unpaid Work and the American Economy," The Economics Review at New York University, published September 1, 2022, https://theeconreview.com/2022/09/01/womens-unpaid-work-and-the-american-economy/.
33. Amanda Ruggieri, "The Stigma of the Stay-At-Home Dad," British Broadcasting Corporation, published February 9, 2023, https://www.bbc.com/worklife/article/20230206-are-stay-at-home-dads-becoming-more-common#:~:text=But%20they're%20still%20relatively,fathers%20and%20non%2Dworking%20mothers.

Bibliography

Adcock, Flyn, David Anderson, and Parr Rosson. "The Economic Impacts of Immigrant Labor on U.S. Dairy Farms." Department of Agricultural Economics, Texas A&M University. Published August 2015. https://www.nmpf.org/wp-content/uploads/immigration-survey-090915.pdf.

Agriculture Fairness Alliance. "2020 USDA Farm Spending." Accessed July 4, 2023. https://agriculturefairnessalliance.org/news/2020-usda-farm-program-spending/#easy-footnote-bottom-2-9183.

Agrifarming. "Poultry Feed Chart and Weight Chart." Last modified August 30, 2023. https://www.agrifarming.in/poultry-feed-chart-and-weight-chart.

Akhtar, Khalil. "Male Chick Culling in Egg Industry Could End, Thanks to Canadian Technology." CBC News. Last modified June 22, 2016. https://www.cbc.ca/news/science/male-chick-culling-technology-1.3647228.

Alt, Kurt W., Ali Al-Ahmad, and Johan Peter Woelber. "Nutrition and Health in Human Evolution—Past to Present." *Nutrients* 14, no. 17 (August 2022): 3594. https://doi.org/10.3390/nu14173594.

American Association for the Advancement of Science—Science Daily. "Diet and Disease in Cattle: High-Grain Feed May Promote Illness and Harmful Bacteria." Published May 11, 2001. https://www.sciencedaily.com/releases/2001/05/010511074623.htm.

American Veterinary Medical Association. "Welfare Implications of Laying Hen Housing." Published January 26, 2012. https://www.avma.org/resources-tools/literature-reviews/welfare-implications-laying-hen-housing.

AnAge: The Animal Ageing and Longevity Database. "AnAge Entry for *Gallus Gallus.*" Accessed May 31, 2023. https://genomics.senescence.info/species/entry.php?species=Gallus_gallus.

Anderson, Dave, Chris Martin, and Andrew Parrella. "Three Cool Facts About Turkeys," Society for the Protection of New Hampshire Forests. Published June 9, 2016. https://forestsociety.org/something-wild/three-cool-facts-about-turkeys#:~:text=A%20turkey%20hen%20lays%20one,to%20lay%20the%20entire%20clutch.

Animal Welfare Institute. "Humane Methods of Slaughter Act." Accessed July 3, 2023. https://awionline.org/content/humane-methods-slaughter-act.

Animal Welfare Institute. "Humane Slaughter Update." Published April 2017. https://awionline.org/sites/default/files/products/FA-HumaneSlaughterReport-2017.pdf.

Animal Welfare Institute. "Inhumane Practices on Factory Farms." Published 2023. https://awionline.org/content/inhumane-practices-factory-farms.

Anthis, Jacy Reese. "US Factory Farming Estimates." Sentience Institute. Last modified April 11, 2019. https://www.sentienceinstitute.org/us-factory-farming-estimates.

Astley, Mark. "Dairy Farmers of America (DFA) Has Agreed to Pay $158.6M to Settle 2007 Lawsuit That Alleged It Conspired with Dean Foods and Others to Suppress Raw Milk Prices across 14 Southeast US States." Dairy Reporter. Last modified January 23, 2013. https://www.dairyreporter.com/Article/2013/01/23/Dairy-Farmers-of-America-agrees-to-156.8m-price-fixing-settlement.

Atalah, Javier, and Pablo Sanchez-Jerez. "Global Assessment of Ecological Risks Associated with Farmed Fish Escapes." *Global Ecology and Conservation* 21 (March 2020). https://doi.org/10.1016/j.gecco.2019.e00842.

Aune, Dagfinn, Deborah A. Navarro Rosenblatt, Doris S. M. Chan, Ana Rita Vieira, Rui Vieira, Darren C. Greenwood, Lars J. Vatten, and Teresa Norat. "Dairy Products, Calcium, and Prostate Cancer Risk: A Systematic Review and Meta-Analysis of Cohort Studies." *The American Journal of Clinical Nutrition* 101, no. 1 (2015): 87–117. https://doi.org/10.3945/ajcn.113.067157.

Ballotpedia. "Tom Vilsack." Accessed July 4, 2023. https://ballotpedia.org/Tom_Vilsack.

Barnard, Neal, Susan Levin, and Caroline Trapp. "Meat Consumption as a Risk Factor for Type 2 Diabetes." *Nutrients* 6, no. 2 (2014): 897–910. https://doi.org/10.3390/nu6020897.

Baron, Eder. "Importance of Feed Withdrawal and the Loading Process to the Poultry Industry." WattPoultry. Published June 1, 2020. https://www.wattagnet.com/articles/40432-importance-of-feed-withdrawal-and-the-loading-process-to-the-poultry-industry.

Bar-On, Yinon M., Rob Phillips, and Ron Milo. "The Biomass Distribution on Earth." *Proceedings of the National Academy of Sciences (PNAS)* 115, no. 25 (May 21, 2018): 6506–11. https://doi.org/10.1073/pnas.1711842115.

Basevich Singer, Isaac. "The Letter Writer." *The New Yorker*. January 13, 1968. https://archives.newyorker.com/newyorker/1968-01-13/flipbook/026/.

Baxter, Emma M., Inger Lise Andersen, and Sandra A. Edwards. "2—Sow Welfare in the Farrowing Crate and Alternatives." In *Advances in Pig Welfare*, edited by Marek Špinka, 27–72. Sawston, UK: Woodhead Publishing, 2018. https://doi.org/10.1016/B978-0-08-101012-9.00002-2.

Beaver, Bonnie, and Gail Golab. "Livestock Welfare Issues." In *The Veterinarian's Guide to Animal Welfare*, edited by Bonnie Beaver and Gail Golab. Cambridge, MA: Academic Press, 2023. https://www.sciencedirect.com/topics/agricultural-and-biological-sciences/dairy-calves.

Beef Quality Assurance. "Beef Quality Assurance Supplemental Guidelines." Published 2014. https://www.bqa.org/Media/BQA/Docs/supplemental_guidelines_2014.pdf.

Begum, Tammana. "What Is Mass Extinction and Are We Facing a Sixth One?" Natural History Museum. Last modified February 21, 2023. https://www.nhm.ac.uk/discover/what-is-mass-extinction-and-are-we-facing-a-sixth-one.html#:~:text=A%20mass%20extinction%20event%20is,less%20than%202.8%20million%20years.

Berg, Charlotte, and Mohan Raj. "A Review of Different Stunning Methods for Poultry—Animal Welfare Aspects." *Animals* 5, no. 4 (November 2015): 1207–19. https://www.researchgate.net/publication/285383557_A_Review_of_Different_Stunning_Methods_for_Poultry-Animal_Welfare_Aspects_Stunning_Methods_for_Poultry.

Berger, John. *About Looking*. New York: Random House Pantheon Books,1980.

Birdfact. "Can Wild Turkeys Fly? (Height, Speed, Distance + FAQs)." Last modified July 13, 2022. https://birdfact.com/articles/can-wild-turkeys-fly.

Blake, William. "Songs of Innocence–Song 8." *The Complete Poetry & Prose of William Blake*, edited by David V. Erdman, 8–9. New York: Anchor Books Doubleday.

Bolton, Sarah E., and Marina A.G. von Keyserlingk. "The Dispensable Surplus Dairy Calf: Is This Issue a 'Wicked Problem' and Where Do We Go from Here?" *Frontiers in Veterinary Science* 8 (April 14, 2021). https://doi.org/10.3389/fvets.2021.660934.

Borba, Vânis Viera, Aaron Lerner, Torsten Matthias, and Yehuda Shoenfeld. "Bovine Milk Proteins as a Trigger for Autoimmune Diseases: Myth or Reality." *International Journal of Celiac Disease* 8, no. 1 (2020). http://pubs.sciepub.com/ijcd/8/1/3/index.html.

Bramwell, Keith. "Artificial Insemination in Turkeys and Chickens." Merck Veterinary Manual. Last modified October 2022. https://www.merckvetmanual.com/poultry/artificial-insemination/artificial-insemination-in-turkeys-and-chickens#:~:text=In%20most%20commercial%20chicken%20production,breeding%20work%20for%20research%20purposes.

Brown, Culum."Fish Pain: An Inconvenient Truth." *Animal Sentience* 3, no. 32 (2016). https://doi.org/10.51291/2377-7478.1069.

Bureau of Labor Statistics. "Employment Characteristics of Families—2022." Published April 19, 2023. https://www.bls.gov/news.release/pdf/famee.pdf.

Burger, John M. "Replicating Milgram—Would People Still Obey Today?" *American Psychologist* 64, no. 1 (2009): 1–2. https://doi.org/10.1037/a0010932.

Burkholder, JoAnn, Bob Libra, Peter Weyer, Susan Heathcote, Dana Kolpin, Peter S. Thorne, and Michael Wichman. "Impacts of Waste from Concentrated Animal Feeding Operations on Water Quality." *Environmental Health Perspectives* 115, no. 2 (2007): 308–12, https://doi.org/10.1289/ehp.8839.

Butcher, Gary D., and Richard Miles. "Salmonella Control and Molting of Egg-Laying Flocks—Are They Compatible?" Veterinary Medicine—Large Animal Clinical Sciences Department, UF/IFAS Extension. Published February 19, 2019. https://edis.ifas.ufl.edu/publication/VM017.

Butler, Rhett A. "Consequences of Deforestation." Mongabay. Published April 1, 2019. https://rainforests.mongabay.com/09-consequences-of-deforestation.html.

Butler, Virgil. "The Cyberactivist." Published August 31, 2003. http://cyberactivist.blogspot.com/2003/08/inside-mind-of-killer.html.

Calderone, Julia. "You May Want to Think Twice Before Buying Expensive Grass-Fed Beef." Business Insider. Published March 25, 2016. https://www.businessinsider.com/grass-fed-claims-beef-bogus-usda-packaging-2016-2.

Calloway, Claire. "Tyson, JBS, and Other Packers Pay Millions to Settle Poultry and Pork Price-Fixing Suits." Food & Power. Published February 4, 2021. https://www.foodandpower.net/latest/title-of-post-yla6k-lmlgr.

Carter, C. Sue, and Stephen W. Porges. "The Biochemistry of Love: An Oxytocin Hypothesis." *EMBO Reports* 14, no. 1 (2012): 12–16. https://doi.org/10.1038/embor.2012.191.

Carter, John, Allison Jones, Mary O'Brien, Jonathan Ratner, and George Wuerthner. "Holistic Management: Misinformation on the Science of Grazed Ecosystems." *International Journal of Biodiversity* 2014, no. 163431 (April 23, 2014). https://doi.org/10.1155/2014/163431.

Centers for Disease Control and Prevention. "Antibiotic Resistance Threats in the United States." Last modified December 2019. https://www.cdc.gov/drugresistance/biggest-threats.html.

Chatham House. "Food System Impacts on Biodiversity Loss." Published February 3, 2021. https://www.chathamhouse.org/2021/02/food-system-impacts-biodiversity-loss/summary.

Chen, Pei-Ju, Hiroko Awata, Atsuko Matsushita, En-Cheng Yang, Kentaro, and Arikawa. "Extreme Spectral Richness in the Eye of the Common Bluebottle Butterfly, Graphium Sarpedon," *Frontiers in Ecology and Evolution* 4, no. 18 (March 2016): 1–12. https://doi.org/10.3389/fevo.2016.00018.

ChickenMag Poultry & Livestock Magazine. "Red Junglefowl—Meet the Ancestor of the Domesticated Chicken." Accessed May 31, 2023. https://chickenmag.com/red-junglefowl/.

ClimateChange.ie. "New Rebuttal to the Myth 'Holistic Management Can Reverse Climate Change.'" Published March 18, 2019. http://www.climatechange.ie/new-rebuttal-to-the-myth-holistic-management-can-reverse-climate-change/.

CO2-Earth Numbers for Living on Earth. "Daily CO_2." Accessed April 23, 2023. https://www.co2.earth/daily-co2#:~:text=420.66%20ppm&text=This%20table%20presents%20the%20most,%3D%20Mauna%20Loa%20Observatory%2C%20Hawaii.

Cojti, Avexnim, Cesar Gomez, and Bia'ni Madsa' Juárez López. "United in Tradition as Peoples of the Corn." Cultural Survival. Published September 4, 2019. https://www.culturalsurvival.org/publications/cultural-survival-quarterly/united-tradition-peoples-corn.

Coleman, A.D. *Looking at Photographs: Animals.* San Francisco: Chronicle Books, 1994.

Comparably. "Slaughterhouse Butcher Salary." Accessed June 7, 2023. https://www.comparably.com/salaries/salaries-for-slaughterhouse-butcher.

Concise Encyclopedia of Business Ethics. "Corporate Personhood." Accessed June 8, 2023. https://conciseencyclopedia.org/entries/corporate-personhood/.

Connecticut Department of Environmental Protection Wildlife Division. "Wild Turkey Fact Sheet." Accessed July 4, 2023. https://portal.ct.gov/-/media/DEEP/wildlife/pdf_files/outreach/fact_sheets/wldturkypdf.pdf.

Consumer Reports. "Why Grass-Fed Beef Costs More." Published August 24, 2015. https://www.consumerreports.org/cro/magazine/2015/08/why-grass-fed-beef-costs-more/index.htm.

Dall, Chris. "FDA: Antibiotic Use in Food Animals Continues to Rise." University of Minnesota Center for Infectious Disease Research Policy. Published December 22, 2016. https://www.cidrap.umn.edu/antimicrobial-stewardship/fda-antibiotic-use-food-animals-continues-rise.

De Natale, Claudia, Giovanni Annuzzi, Lutgarda Bozzetto, Raffaella Mazzarella, Giuseppina Costabile, Ornella Ciano, Gabriele Riccardi, and Angela A. Rivellese. "Effects of a Plant-Based High-Carbohydrate/High-Fiber Diet versus High-Monounsaturated Fat/Low-Carbohydrate Diet on Postprandial Lipids in Type 2 Diabetic Patients." *Diabetes Care* 32, no. 12 (2009): 2168–73. https://doi.org/10.2337/dc09-0266.

Denchak, Melissa. "Greenhouse Effect 101." National Resources Defense Council. Published June 5, 2023. https://www.nrdc.org/stories/greenhouse-effect-101#intro.

De Vries, A., and M.I. Marcondes. "Review: Overview of Factors Affecting Productive Lifespan of Dairy Cows." *Animal* 14, no. S1 (2020): s155–s164. https://doi.org/10.1017/S1751731119003264.

Dhanwal, Dinesh K., Elaine M. Dennison, Nick C. Harvey, and Cyrus Cooper. "Epidemiology of Hip Fracture: Worldwide Geographic Variation." *Indian Journal of Orthopaedics* 45, no. 1 (2011): 15–22. https://doi.org/10.4103/0019-5413.73656.

Dolstein, Josefin. "Holocaust Survivor Likens Treatment of Livestock to Shoah." *The Times of Israel.* October 7, 2016. https://www.timesofisrael.com/holocaust-survivor-likens-treatment-of-livestock-to-shoah/.

Drouillard, James S. "Current Situation and Future Trends for Beef Production in the United States of America—A Review." *Asian-Australasian Journal of Animal Sciences* 31, no. 7 (July 2018): 1007–16. https://doi.org/10.5713/ajas.18.0428.

Duarte, Carlos M., Lucille Chapuis, Shaun P. Collin, Daniel P. Costa, Reny P. Devassy, Victor M. Eguiluz, Christine Erbe, et al. "The Soundscape of the Anthropocene Ocean." *Science* 371, no. 6529 (February 5, 2021). https://doi.org/10.1126/science.aba4658.

Dubner, Stephen J. "Artificial Insemination: What About the Other Animals?" Freakonomics. Published November 22, 2011. https://freakonomics.com/2011/11/artificial-insemination-what-about-the-other-animals/.

Ducks Unlimited. "Amazing Waterfowl Facts." Accessed July 4, 2023. https://www.ducks.org/conservation/waterfowl-research-science/amazing-waterfowl-facts#:~:text=Most%20waterfowl%20fly%20at%20speeds,during%20an%20eight%2Dhour%20flight.

Earthjustice. "Over Fifty Groups Petition to Improve Oversight of Water Pollution for Concentrated Animal Feeding Operations." Published October 26, 2022. https://earthjustice.org/press/2022/over-fifty-groups-petition-epa-to-improve-oversight-of-water-pollution-from-concentrated-animal-feeding.

The Editors of Encyclopedia Britannica. "Chicken." *Encyclopedia Britannica.* Last modified May 18, 2023. https://www.britannica.com/animal/chicken.

Eisen, Michael B., and Patrick O. Brown. "Rapid Global Phaseout of Animal Agriculture Has the Potential to Stabilize Greenhouse Gas Levels for Thirty Years and Offset 68% Of CO_2 Emissions This Century." *Public Library of Science Climate* 1, no. 2, (February 1, 2022). https://doi.org/10.1371/journal.pclm.0000010.

Encyclopedia Britannica, "Livestock Slaughter Procedures." Accessed June 12, 2023. https://www.britannica.com/technology/meat-processing/Livestock-slaughter-procedures.

Environmental Working Group. "PCBs in Farmed Salmon." Published July 31, 2003. https://www.ewg.org/research/pcbs-farmed-salmon.

Espinosa-Marrón, Alan, Kate Adams, Lea Sinno, Alejandra Cantu-Aldana, Martha Tamez, Abrania Marrero, Shilpa N. Bhupathiraju, and Josiemer Mattei. "Environmental Impact of Animal-Based Food Production and the Feasibility of a Shift Toward Sustainable Plant-Based Diets in the United States." *Frontiers in Sustainability* 3, no. 841106 (May 22). https://doi.org/10.3389/frsus.2022.841106.

European Society of Cardiology. "Study Strengthens Links Between Red Meat and Heart Disease." Published April 15, 2021. https://www.escardio.org/The-ESC/Press-Office/Press-releases/study-strengthens-links-between-red-meat-and-heart-disease.

Fairr A Coller Initiative. "Gestation Crates: A Growing Financial Risk." Published February 22, 2022. https://www.fairr.org/article/gestation-crates/.

Farah, Troy. "U.S. Rivers and Lakes Are Shrinking for a Surprising Reason: Cows." *The Guardian*, July 2, 2020. https://www.theguardian.com/environment/2020/jul/02/agriculture-cattle-us-water-shortages-colorado-river.

Favre, David. "Overview of U.S. Animal Welfare Act." Michigan State University College of Law Animal Legal Historical Center. Published 2002. https://www.animallaw.info/article/overview-us-animal-welfare-act.

Federal Bureau of Prisons. "Inmate Gender." Last modified June 24, 2023. https://www.bop.gov/about/statistics/statistics_inmate_gender.jsp.

Federal Register. "Modernization of Swine Slaughter Inspection." Published October 1, 2019. https://www.federalregister.gov/documents/2019/10/01/2019-20245/modernization-of-swine-slaughter-inspection.

Fernandez-Cornejo, Jorge, Richard Nehring, Craig Osteen, Seth Wechsler, Andrew Martin, Alex Vialou. "Pesticide Use in U.S. Agriculture: Twenty-One Selected Crops, 1960–2008." USDA Economic Research Service Economic Information Bulletin Number 124. Published May 2014. https://www.ers.usda.gov/webdocs/publications/43854/46734_eib124.pdf.

Fishcount.org.uk. "Number of Farmed Fish Slaughtered Each Year. Last modified 2019. http://fishcount.org.uk/fish-count-estimates-2/numbers-of-farmed-fish-slaughtered-each-year.

Fishcount.org.uk. "Number of Fish Caught from the Wild Each Year." Last modified 2019. http://fishcount.org.uk/fish-count-estimates-2/numbers-of-fish-caught-from-the-wild-each-year.

Fitzgerald, Amy J., Linda Kalof, and Thomas Dietz. "Slaughterhouses and Increased Crime Rates: An Empirical Analysis of the Spillover From 'The Jungle' Into the Surrounding Community." *Organization & Environment* 22, no. 2 (2009): 158–84. https://doi.org/10.1177/1086026609338164.

Food Chain Workers Alliance. "Hands That Feeds Us." Published June 1, 2012. https://foodchainworkers.org/2012/06/the-hands-that-feed-us/; http://foodchainworkers.org/wp-content/uploads/2012/06/Hands-That-Feed-Us-Report.pdf.

Foodprint. "Pesticides in Our Food System." Last modified February 8, 2021. https://foodprint.org/issues/pesticides/.

FoodPrint. "What Happens to Animal Waste?" Last modified May 8, 2023. https://foodprint.org/issues/what-happens-to-animal-waste/#easy-footnote-bottom-1-1324.

Fox, Carly, Rebecca Fuentes, Fabiola Ortiz Valdez, Gretchen Purser, and Kathleen Sexsmith. "Milked: Immigrant Dairy Farmworkers in New York State." Workers' Center of Central New York and the Worker Justice Center of New York. Published 2017. https://milkedny.files.wordpress.com/2017/05/milked_053017.pdf.

Francis, John Michael, ed. "Columbian Exchange—Livestock." In *Iberia and the Americas: Culture, Politics, and History: A Multidisciplinary Encyclopedia*. Santa Barbara, CA: ABC-CLIO, 2006.

Frank, John W., Ronald S. Moore, and Genevieve M. Ames. "Historical and Cultural Roots of Drinking Problems Among American Indians." *American Journal of Public Health* 90, no. 3 (2000): 344–51. https://doi.org/10.2105/ajph.90.3.344.

Frantz, Douglas, and Catherine Collins. "Three Reasons to Avoid Farmed Salmon." *Time*, July 21, 2022. https://time.com/6199237/is-farmed-salmon-healthy-sustainable/.

Fraser, Gary E., Karen Jaceldo-Siegl, Michael Orlich, Andrew Mashchak, Rawiwan Sirirat, and Synnove Knutsen. "Dairy, Soy, and Risk of Breast Cancer: Those Confounded Milks." *International Journal of Epidemiology* 49, no. 5 (October 2020): 1526–37 https://doi.org/10.1093/ije/dyaa007.

Fredrickson, Torgny N. "Ovarian Tumors of the Hen." *Environmental Health Perspectives* 73 (August 1987): 35–51. https://doi.org/10.1289/ehp.877335.

Fremstad, Shawn, Hye Jin Rho, and Hayley Brown. "Meatpacking Workers Are a Diverse Group Who Need Better Protections." Center for Economic and Policy Research. Published April 29, 2020. https://cepr.net/meatpacking-workers-are-a-diverse-group-who-need-better-protections/.

Gaard, Greta. "Living Interconnections with Animals and Nature." In *Ecofeminism Women, Animals, Nature*, edited by Greta Gaard, 5. Philadelphia: Temple University Press, 1993.

Gaarder, Emily. *Women and the Animal Rights Movement*. New Brunswick, NJ: Rutgers University Press, 2011.

Gaeng, Jennifer. "Duck Lifespan: How Long Do Ducks Live?" Last modified June 27, 2023. https://a-z-animals.com/blog/duck-lifespan-how-long-do-ducks-live/.

Ghirlanda, Stefano, Lisolette Jansson, and Magnus Enquist. "Chickens Prefer Beautiful Humans." *Human Nature* 13 (2002): 383–9. https://doi.org/10.1007/s12110-002-1021-6.

Gibbs, Joshua, and Franceso P. Cappuccio. "Plant-Based Dietary Patterns for Human and Planetary Health." *Nutrients* 14, no. 8 (2022): 1614. https://doi.org/10.3390/nu14081614.

Gonçalves, Juliano L., Juliana L. de Campos, Andrew J. Steinberger, Nasia Safdar, Ashley Kates, Ajay Sethi, John Shutske, Garret Suen, Tony Goldberg, Roger I. Cue, andPamela L. Ruegg. "Incidence and Treatments of Bovine Mastitis and Other Diseases on Thirty-Seven Dairy Farms in Wisconsin." *Pathogens* 11, no. 11 (2022): 1282. https://doi.org/10.3390/pathogens11111282.

Goodland, Robert, and Jeff Anhang. "Livestock and Climate Change: What If the Key Actors in Climate Change Are Cows, Pigs, and Chickens?" *World Watch* (November/December 2009): 10–19. https://awellfedworld.org/wp-content/uploads/Livestock-Climate-Change-Anhang-Goodland.pdf.

Gordon, Ethan, Federico Davila, and Chris Riedy. "Transforming Landscapes and Mindscapes Through Regenerative Agriculture." *Agriculture and Human Values* 39 (2022): 809–26. https://doi.org/10.1007/s10460-021-10276-0.

Graham, Jay P., and Keeve E. Nachman. "Managing Waste from Confined Animal Feeding Operations in the United States: The Need for Sanitary Reform." *Journal of Water Health* 8, no. 4 (2010): 646–70. https://doi.org/10.2166/wh.2010.075.

Granquist, E.G., G. Vasdal, I.C. de Jong, and R.O. Moe. "Lameness and Its Relationship with Health and Production Measures in Broiler Chickens." *Animal* 13, no. 10 (2019): 2365–72. https://doi.org/10.1017/S1751731119000466.

A Greener World. "A Greener World Technical Advice Fact Sheet No. 13 The Prolific Hybrid Hen: Why Less Is More When It Comes to Egg Laying." Published 2009. https://agreenerworld.org/wp-content/uploads/2020/05/TAFS-13-The-Prolific-Hybrid-Hen-v4.pdf.

Handwerk, Brian. "An Evolutionary Timeline of *Homo Sapiens*." *Smithsonian Magazine*. February 2, 2021. https://www.smithsonianmag.com/science-nature/essential-timeline-understanding-evolution-homo-sapiens-180976807/.

Harris, Chris. "Poultry Welfare During Transport." The Poultry Site. Published September 3, 2015. https://www.thepoultrysite.com/articles/poultry-welfare-during-transport.

Hayek, Matthew N., and Rachael D. Garrett. "Nationwide Shift to Grass-Fed Beef Requires Larger Cattle Population." *Environmental Research Letters* 13, no. 8 (July 25, 2018). https://doi.org/10.1088/1748-9326/aad401.

Hayes, Jared. "USDA Livestock Subsidies near $50 Billion, EWG Analysis." Environmental Working Group. Published February 28, 2022. https://www.ewg.org/news-insights/news/2022/02/usda-livestock-subsidies-near-50-billion-ewg-analysis-finds.

Heim, Tracy. "The Indigenous Origins of Regenerative Agriculture." National Farmers Union. Published October 12, 2020. https://nfu.org/2020/10/12/the-indigenous-origins-of-regenerative-agriculture/.

Held, Lisa. "Just a Few Companies Control the Meat Industry. Can a New Approach to Monopolies Level the Playing Field?" Civil Eats. Published July 14, 2021. https://civileats.com/2021/07/14/just-a-few-companies-control-the-meat-industry-can-a-new-approach-to-monopolies-level-the-playing-field/.

Herzog, Harold A. "Gender Differences in Human-Animal Interactions: A Review." *Anthrozoos* 20, no. 1 (2007). https://cdn2.psychologytoday.com/assets/attachments/44240/gender-and-human-animal-interactions.pdf.

Hites, Ronald A., Jeffery A. Foran, David O. Carpenter, M. Coreen Hamilton, Barbara A. Knuth, and Steven J. Schwager. "Global Assessment of Organic Contaminants in Farmed Salmon." *Science* 303, no. 5655 (2004): 226–9. https://doi.org/10.1126/science.1091447.

Hofste, Rutger Willem, Paul Reig, and Leah Schleifer. "Seventeen Countries, Home to One-Quarter of the World's Population, Face Extremely High Water Stress." World Resources Institute. Published August 6, 2019. https://www.wri.org/insights/17-countries-home-one-quarter-worlds-population-face-extremely-high-water-stress.

Holmes, Matilda, Helena Hamerow, and Richard Thomas. "Close Companions? A Zooarchaeological Study of the Human-Cattle Relationship in Medieval England." *Animals (Basel)* 11, no. 4 (April 20, 2021). https://doi.org/10.3390/ani11041174.

Howell, Cheryl. "Egg Binding in Birds—The Symptoms, Causes, and Treatment?" The Happy Chicken Coop. Published December 6, 2022. https://www.thehappychickencoop.com/egg-binding-in-birds/#:~:text=to%20lay%20normally.-,Overproduction%20of%20Eggs,in%20egg%20binding%20in%20birds.

The Humane Society of the United States. "Cage-free vs. Battery-Cage Eggs." Accessed July 3, 2023. https://www.humanesociety.org/resources/cage-free-vs-battery-cage-eggs#:~:text=Cage%2DFree%20Systems&text=Unlike%20battery%20hens%2C%20cage%2Dfree,hens%20who%20never%20go%20outside.

Husch Blackwell. "State-By-State Guide to Ag-Gag Legislation." Last modified May 22, 2023. https://www.huschblackwell.com/newsandinsights/state-by-state-guide-to-ag-gag-legislation#:~:text=Several%20states%20have%20made%20attempts,Amendment's%20guarantee%20of%20free%20speech.

International Chicken Genome Sequencing Consortium. "Sequence and Comparative Analysis of the Chicken Genome Provide Unique Perspectives on Vertebrate Evolution." *Nature* 432 (2004): 695–716. https://doi.org/10.1038/nature03154.

International Union for Conservation of Nature. "Biodiversity and Soil Health: How Protecting One, Safeguards the Other." Published May 19, 2022. https://www.iucn.org/news/water/202205/biodiversity-and-soil-health-how-protecting-one-safeguards-other.

International Union for Conservation of Nature. "Marine Plastic Pollution." Published November 2021. https://www.iucn.org/resources/issues-brief/marine-plastic-pollution#:~:text=Over%20400%20million%20tons%20of,waters%20to%20deep%2Dsea%20sediments.

Iowa Department of Natural Resources. "Animals to Animal Units Conversion." Published January 2021. https://www.iowadnr.gov/Portals/idnr/uploads/forms/5420020.pdf.

Jolley, Chuck. "What's Behind the New Swine Slaughter Inspection System?" Food Safety News. Published January 10, 2020. https://www.foodsafetynews.com/2020/01/whats-behind-the-new-swine-slaughter-inspection-system/.

Jones, Dena. "The Welfare of Birds at Slaughter in the United States." Animal Welfare Institute. Published November 2020. https://awionline.org/sites/default/files/publication/digital_download/20TheWelfareBirdsSlaughter.pdf.

Jones, Erin, Ariane Datil, and Alanna Delfino. "Yes, Cattle Are the Top Source of Methane Emissions in the U.S." Verify. Last modified December 14, 2021. https://www.verifythis.com/article/news/verify/environment-verify/cattle-cows-the-top-source-of-methane-emissions-in-united-states/536-8d5bf326-6955-4a9c-8ea5-761d73ba464c.

Jones, Nicola. "How the World Passed a Carbon Threshold and Why It Matters." Yale Environment 360. Published January 26, 2017. https://e360.yale.edu/features/how-the-world-passed-a-carbon-threshold-400ppm-and-why-it-matters.

Juhl, Christian R., Helle K. M. Bergholdt, Iben M. Miller, Gregor B. E. Jemec, Jørgen K. Kanters, and Christina Ellervik. "Dairy Intake and Acne Vulgaris: A Systematic Review and Meta-Analysis of 78,529 Children, Adolescents, and Young Adults." *Nutrients* 10, no. 8 (August 9, 2018): 1049. https://doi.org/10.3390/nu10081049.

Kahleova, Hana, Kitt Falk Petersen, Gerald I. Shulman, Jihad Alwarith, Emilie Rembert, Andrea Tura, Martin Hill, Richard Holubkov, and Neal D. Barnard. "Effect of a Low-Fat Vegan Diet on Body Weight, Insulin Sensitivity, Postprandial Metabolism, and Intramyocellular and Hepatocellular Lipid Levels in Overweight Adults." *JAMA Network Open* 3, no. 11 (2020): e2025454. https://doi.org/10.1001/jamanetworkopen.2020.25454.

Kasaeinasab, Abbasali, Mehdi Jahangiri, Ali Karimi, Hamid Reza Tabatabaei, and Sonia Safari. "Respiratory Disorders Among Workers in Slaughterhouses." *Safety and Health at Work* 8, no. 1 (2017): 84–8. https://doi.org/10.1016/j.shaw.2016.04.002.

Kaufman, Dan. "Is It Time to Break Up Big Ag?" *The New Yorker*, April 17, 2021. https://www.newyorker.com/news/dispatch/is-it-time-to-break-up-big-ag.

Kemmerer, Lisa. "Evidence of Sexism and Male Privilege in the Animal Liberation/Rights Movement." *Between the Species* 21, no. 1 (Spring 2018). http://lisakemmerer.com/articles/evidence_of_sexism_and_male_privilege_in_the_animal_liberation_rights_movement.pdf.

Kemmerer, Lisa. "The 'Why' of Sexism in Social Justice Movements." In *Oppressive Liberation*, 69–93. London: Palgrave Macmillan, 2023. https://doi.org/10.1007/978-3-031-15363-1_4.

Ketcham, Christopher. "Allan Savory's Holistic Management Theory Falls Short on Science." *Sierra Club Magazine*. February 23, 2017. https://www.sierraclub.org/sierra/2017-2-march-april/feature/allan-savory-says-more-cows-land-will-reverse-climate-change.

Ketcham, Christopher. "The Rogue Agency—A USDA Program That Tortures and Kills Endangered Species. *Harpers*. March 2016. https://harpers.org/archive/2016/03/the-rogue-agency/.

Kevany, Sophie. "More than Twenty Million Farm Animals Die on Way to Abattoir in US Every Year." *The Guardian*, June 15, 2022. https://www.theguardian.com/environment/2022/jun/15/more-than-20-million-farm-animals-die-on-way-to-abattoir-in-us-every-year.

Kim, Hyunju, Laura E. Caulfield, Vanessa Garcia-Larsen, Lyn M. Steffen, Josef Coresh, and Casey M. Rebholz. "Plant-Based Diets Are Associated With a Lower Risk of Incident Cardiovascular Disease, Cardiovascular Disease Mortality, and All-Cause Mortality in a General Population of Middle-Aged Adults." *Journal of the American Heart Association* 8, no. 16 (2019). https://www.ahajournals.org/doi/10.1161/JAHA.119.012865.

Kirschbaum, Kari, and Jason McCullough. "*Melagris Gallopavo* Wild Turkey." Animal Diversity Web. Published 2001. https://animaldiversity.org/site/accounts/information/Meleagris_gallopavo.html#6ffa8ae020c377c6478eb5af38571f6a.

Kivi, Rose. "How Does a Turkey Reproduce?" Sciencing. Last modified April 19, 2018. https://sciencing.com/a-turkey-reproduce-4709177.html.

Knolle, Franziska, Rita P. Goncalves, and A. Jennifer Morton. "Sheep Recognize Familiar and Unfamiliar Human Faces from Two-Dimensional Images." *Royal Society of Open Science* 4, no. 11 (2017): 171228. https://doi.org/10.1098/rsos.171228.

Khodayari, Shaghayegh, Omid Sadeghi, Maryam Safabakhsh, Hassan Mozaffari-Khosravi. "Meat Consumption and the Risk of General and Central Obesity: The Shahedieh Study." *BMC Research Notes* 15, no. 339 (2022). https://doi.org/10.1186/s13104-022-06235-5.

Koch, Alexander, Chris Brierley, Mark M. Maslin, and Simon L. Lewis. "Earth System Impacts of the European Arrival and Great Dying in the Americas After 1492." *Quaternary Science Reviews* 207 (2019): 13–36. https://doi.org/10.1016/j.quascirev.2018.12.004.

Kolbe, Emily A. "'Won't You Be My Neighbor?' Living with Concentrated Animal Feeding Operations." *Iowa Law Review* 99, no. 1 (November 15, 2013): 427. https://ilr.law.uiowa.edu/print/volume-99-issue/wont-you-be-my-neighbor-living-with-concentrated-animal-feeding-operations.

Kram, Yoseph A., Stephanie Mantey, and Joseph C. Corbo. "Avian Cone Photoreceptors Tile the Retina as Five Independent, Self-Organizing Mosaics." *PLOS ONE* 5, no. 4 (February 1, 2010). https://doi.org/10.1371/journal.pone.0008992.

Kuhn, Casey. "'Fire Is Medicine': How Indigenous Practices Could Help Curb Wildfires." Public Broadcasting Service News Hour. Published July 23, 2021. https://www.pbs.org/newshour/science/fire-is-medicine-how-indigenous-practices-could-help-curb-wildfires.

Kukura, Jared. "Theoretical Ecological Collapse: Any Excuse to Kill Elephants." *The Journal of African Elephants*. Published April 2, 2020. https://www.africanelephantjournal.com/theoretical-ecological-collapse-any-excuse-to-kill-elephants/.

Lambert, G.W., C. Reid, D.M. Kaye, G.L. Jennings, and M.D. Esler. "Effect of Sunlight and Season on Serotonin Turnover in the Brain." *The Lancet* 360, no. 9348 (December 2002): 1840–42. https://doi.org/10.1016/S0140-6736(02)11737-5.

Lupo, Christopher D., David E. Clay, Jennifer L. Benning, and James J. Stone. "Life-Cycle Assessment of the Beef Cattle Production System for the Northern Great Plains, USA." *Journal of Environmental Quality* 42, no. 5 (September–October 2013). https://doi.org/10.2134/jeq2013.03.0101.

MacDonald, James M., and William D. McBride. "The Transformation of U.S. Livestock Agriculture Scale, Efficiency, and Risks." *Economic Research Service Economic Information Bulletin Number 43* (USDA January 2009). https://www.ers.usda.gov/webdocs/publications/44292/10992_eib43.pdf?v=0.

Macnair, Rachel M. *Perpetration-Induced Traumatic Stress: The Psychological Consequences of Killing*. Westport, CT: Praeger Publishers, 2002. https://books.google.com/books?hl=en&lr=&id=enEOd8YgR0YC&oi=fnd&pg=PP11&dq=perpetration-induced+traumatic+stress+slaughterhouse&ots=9WkMyYNMPa&sig=l1NE9RjCGwyc7m1kJZPUKFGuFio#v=onepage&q&f=false.

Madigan, Mariah, and Elisa Karhu. "The Role of Plant-Based Nutrition in Cancer Prevention." *Journal of Unexplored Medical Data* 3, no. 9 (2018). http://dx.doi.org/10.20517/2572-8180.2018.05.

Maes, Dominiek, Alfonso Lopez Rodriguez, Tom Rijsselaere, Philip Vyt, and Ann Van Soom. "Artificial Insemination in Pigs." In *Artificial Insemination in Farm Animals*, edited by Milad Manafi, Introduction. London: InTech, 2011. https://doi.org/10.5772/16592.

Marino, Lori. "Thinking Chickens: A Review of Cognition, Emotion, and Behavior in the Domestic Chicken." *Animal Cognition* 20, no. 2 (2017): 127–47. https://doi.org/10.1007/s10071-016-1064-4.

Mathur, Shubh. "Dead Zones and Toxic Algae Bloom in U.S. Waters." *Environment & Society Portal, Arcadia*, no. 7 (Spring 2017). https://doi.org/10.5282/rcc/7748).

Maughn, Ralph. "Alan Savory Gives a Popular and Very Misleading Ted Talk." The Wildlife News. Published March 18, 2013. https://www.thewildlifenews.com/2013/03/18/alan-savory-gives-a-popular-and-very-misleading-ted-talk/.

McWilliams, James. "All Sizzle and No Steak." Slate. Published April 22, 2013. https://slate.com/human-interest/2013/04/allan-savorys-ted-talk-is-wrong-and-the-benefits-of-holistic-grazing-have-been-debunked.html.

McWilliams, James. "Milk of Human Kindness." Forbes. October 25, 2013. https://www.forbes.com/sites/jamesmcwilliams/2013/10/25/milk-of-human-kindness-denied-to-dairy-cows/?sh=7dadeae6598b.

Mekonnen, Mesfin M., and Arjen Y. Hoekstra. "A Global Assessment of the Water Footprint of Farm Animal Products." *Ecosystems* 15 (2012): 401–15. https://doi.org/10.1007/s10021-011-9517-8.

Menegat, Stefano, Alicia Ledo, and Reyes Tirado. "Greenhouse Gas Emissions from Global Production and Use of Nitrogen Synthetic Fertilisers in Agriculture." *Scientific Reports* 12, no. 14490 (2022). https://doi.org/10.1038/s41598-022-18773-w.

Micha, Renata, George Michas, and Darius Mozaffarian. "Unprocessed Red and Processed Meats and Risk of Coronary Artery Disease and Type 2 Diabetes—an Updated Review of the Evidence." *Current Atherosclerosis Reports* 14, no. 6 (September 2012): 515–24. https://doi.org/10.1007/s11883-012-0282-8.

Milman, Oliver. "'A Barbaric Federal Program': U.S. Killed 1.75 Million Animals Last Year—Or 200 per Hour." The *Guardian*. March 25, 2022. https://www.theguardian.com/world/2022/mar/25/us-government-wildlife-services-animals-deaths.

Milman, Oliver. "Meat Industry Blamed for Largest Ever 'Dead Zone' in Gulf of Mexico." *The Guardian*. August 1, 2017. https://www.theguardian.com/environment/2017/aug/01/meat-industry-dead-zone-gulf-of-mexico-environment-pollution.

Milne, Sandy. "How Water Shortages Are Brewing Wars." BBC Future. Published August 16, 2021. https://www.bbc.com/future/article/20210816-how-water-shortages-are-brewing-wars.

Mississippi State University Extension. "Economic Impact of Artificial Insemination vs. Natural Mating for Beef Cattle Herds." Publication Number P2486. Last modified 2020. http://extension.msstate.edu/publications/economic-impact-artificial-insemination-vs-natural-mating-for-beef-cattle-herds.

Modern Farmer. "Everything You Want To Know About Turkey Sex." Published November 16, 2016. https://modernfarmer.com/2016/11/everything-want-know-turkey-sex/#:~:text=How%20Wild%20Turkeys%20Mate,)%20or%20jennies%20(females).

Moens, Jonathan. "The Egg Industry Grapples with a Grim Practice: Chick Culling." Undark. Published March 15, 2021. https://undark.org/2021/03/15/egg-industry-grapples-with-chick-culling/.

Monbiot, George. "Eat More Meat and Save the World: The Latest Implausible Farming Miracle." *The Guardian*. August 4, 2014. https://www.theguardian.com/environment/georgemonbiot/2014/aug/04/eat-more-meat-and-save-the-world-the-latest-implausible-farming-miracle.

Mordecai, Gideon. "Fish Farms Transmit Viruses to Endangered Wild Pacific Salmon, New Evidence Shows." The Conversation. Published May 26, 2021. https://theconversation.com/fish-farms-transmit-viruses-to-endangered-wild-pacific-salmon-new-evidence-shows-158691.

Mt. Pleasant, Jane. "Food Yields and Nutrient Analyses of the Three Sisters: A Haudenosaunee Cropping System." *Ethnobiology Letters* 7, no. 1 (2016): 87–98. https://doi.org/10.14237/ebl.7.1.2016.721.

National Human Genome Research Institute. "Researchers Compare Chicken, Human Genomes." Published December 8, 2004. https://www.genome.gov/12514316/2004-release-researchers-compare-chicken-human-genomes.

National Institutes of Health National Library of Medicine. "AD 1493–1550s: Native Peoples Begin Dying from European Diseases." Accessed July 4, 2023. https://www.nlm.nih.gov/nativevoices/timeline/169.html.

National Poultry Council. "Per Capita Consumption of Poultry and Livestock, 1965 to Forecast 2022, in Pounds." Last modified December 2021. https://www.nationalchickencouncil.org/about-the-industry/statistics/per-capita-consumption-of-poultry-and-livestock-1965-to-estimated-2012-in-pounds/.

National Wild Turkey Federation. "Wild Turkey Status Update." Published November 18, 2022. https://www.nwtf.org/content-hub/wild-turkey-status-update.

New York State Department of Environmental Conservation. "Wild Turkey Habitat Management." Accessed June 5, 2023. https://www.dec.ny.gov/animals/7288.html#:~:text=Home%20Range&text=A%20turkey%20can%20and%20does,annual%20needs%20on%20your%20land.

Ngatia, Lucy, Johnny M. Grace III, Daniel Moriasi, and Robert Taylor. "Nitrogen and Phosphorous Eutrophication in Marine Ecosystems," in *Monitoring of Marine Pollution*, edited by Houma Bachari Fouzia. London: IntechOpen Limited, 2019. https://doi.org/10.5772/intechopen.81869.

Nicol, Christine. "Feather Pecking and Cannibalism." In *Advances in Poultry Welfare*, edited by Joy A. Mench, 175–97. Duxford, UK: Woodhead Publishing, 2018. https://www.sciencedirect.com/topics/agricultural-and-biological-sciences/debeaking.

Nordborg, Maria, *Holistic Management—A Critical Review of Allan Savory's Grazing Method.* Uppsala, Sweden: SLU/EPOK—Centre for Organic Food & Farming & Chalmers, 2016. https://orgprints.org/id/eprint/34330/1/holisticmanagement_review.pdf.

O'Connor, Michael L., and Jana L. Peters. "Artificial Insemination Technique: Cattle." Last modified May 13, 2022. https://extension.psu.edu/artificial-insemination-technique-cattle.

Office of the Law Revision Counsel. "U.S. Congress, Code § 80502—Transportation of Animals." Published July 5, 1994. https://uscode.house.gov/view.xhtml?req=granuleid:USC-prelim-title49-section80502&num=0&edition=prelim#sourcecredit.

Onder, Damla. "Women's Unpaid Work and the American Economy." The Economics Review at New York University. Published September 1, 2022. https://theeconreview.com/2022/09/01/womens-unpaid-work-and-the-american-economy/.

Open Secrets. "Dairy: Lobbying." Accessed July 4, 2023. https://www.opensecrets.org/industries/lobbying.php?cycle=2022&ind=A04.

Open Secrets. "Lobbying." Accessed July 4, 2023. https://www.opensecrets.org/industries/lobbying.php?cycle=2022&ind=A04.

Open Secrets. "Meat Processing & Products: Lobbying." Accessed July 4, 2023. https://www.opensecrets.org/industries/lobbying.php?ind=G2300.

Orgiazzi, Alberto, and Panos Panagos. "Soil Biodiversity and Soil Erosion: It Is Time to Get Married." *Global Ecology and Biogeogeography* 27, no. 10 (2018): 1155–67. https://doi.org/10.1111/geb.12782.

Pachirat, Timothy. *Every Twelve Seconds.* New Haven: Yale University Press, 2013.

Pan, An, Qi Sun, Adam M. Bernstein, Matthias B. Schulze, JoAnn E. Manson, Walter C. Willett, and Frank B. Hu. "Red Meat Consumption and Risk of Type 2 Diabetes: 3 Cohorts of US Adults and an Updated Meta-Analysis." *The American Journal of Clinical Nutrition* 94, no. 4 (October 2011): 1088–96. https://doi.org/10.3945/ajcn.111.018978.

Panko, Ben. "Thousands of Invasive Salmon Escape From Farm in Pacific Northwest." *Smithsonian Magazine*, August 24, 2017. https://www.smithsonianmag.com/smart-news/thousands-invasive-salmon-escape-farm-pacific-northwest-180964619/.

Papscun, Dan. "Pilgrim's Ex-Execs' Charges Dropped in End to DOJ Chicken Probe (2)." Bloomberg Law. Last modified October 17, 2022. https://news.bloomberglaw.com/antitrust/ex-pilgrims-execs-get-antitrust-charges-dropped-in-chicken-case.

Park, Sunmin, Nobuko Hongu, and James W. Daily III. "Native American Foods: History, Culture, and Influence on Modern Diets." *Journal of Ethnic Foods* 3, no. 3 (September 2016): 171–7. https://doi.org/10.1016/j.jef.2016.08.001.

Pawlinger, Michelle, and Dena Jones. "Animals in Transport Languish as Twenty-Eight Hour Law Goes Off the Rails." *Lewis & Clark School Animal Law Review* 25, no. 1 (2018). https://law.lclark.edu/law_reviews/animal_law_review/past_issues/volume_25_1.php.

Penn State Extension. "Grass-fed Beef Production." Last modified March 7, 2018. https://extension.psu.edu/grass-fed-beef-production.

Penn State Extension. "Grass Fed vs. Grain Fed Beef." Accessed May 3, 2023. https://extension.psu.edu/grass-fed-vs-grain-fed-beef.

Penn State Extension. "Modern Turkey Industry." Accessed July 4, 2023. https://extension.psu.edu/programs/4-h/opportunities/projects/animal-science/poultry/raising-rearing/viii-other-online-resources/the-chicken/modern-turkey-industry.

Peterson, Chris. "Pigs Go Insane in Gestation Crates." *The Asbury Park Press*, November 26, 2014. https://www.app.com/story/opinion/columnists/2014/11/26/petersen-pigs-go-insane-gestation-crates/19542201/.

Physicians Committee for Responsible Medicine. "Dairy Consumption Linked to Prostate, Ovarian, Breast Cancers, Finds New Narrative Review." Published March 7, 2023. https://www.pcrm.org/news/news-releases/dairy-consumption-linked-prostate-ovarian-breast-cancers-finds-new-narrative#:~:text=The%20Adventist%20Health%20Study%2D2,fat%20milks%20produced%20similar%20results.

Pitney, Nico. "Scientists Believe the Chickens We Eat Are Being Slaughtered While Conscious." Huffington Post. Published October 28, 2016. www.huffpost.com/entry/chickens-slaughtered-conscious_n_580e3d35e4b000d0b157bf98.

Poinski, Megan. "Meat Companies Conspired to Set Wages Low, Lawsuit Claims." Divewire by Industry Dive. Published November 23, 2022. https://www.fooddive.com/news/meat-wage-lawsuit-filed/637103/.

Pollmann, D. Steven. "Sow Lifetime Productivity: Importance of Monitoring in Commercial Pork Production." *Allen D. Leman Swine Conference* 39 (2012). https://conservancy.umn.edu/bitstream/handle/11299/139636/Pollmann.pdf?sequence=1&isAllowed=y.

Poore, Joseph, and Thomas Nemecek. "Reducing Food's Environmental Impacts through Producers and Consumers." *Science* 360, no. 6392 (2018): 987–92. https://doi.org/10.1126/science.aaq0216.

Pye, Alison, Kristin Bash, Adam Joiner, and Jane Beenstock. "Good for the Planet and Good for Our Health: the Evidence for Whole-Food Plant-Based Diets." *BJ Psychology International* 19, no. 4 (2022): 90–2. https://doi.org/10.1192/bji.2022.7.

Rainey, Malcolm, and Paul Vining. "Artificial Insemination in Swine: Breeding the Female." Published June 13, 2022. https://www.aces.edu/blog/topics/farming/artificial-insemination-in-swine-breeding-the-female/.

Rajaram, Govindarajan, Donald C. Erbach, and Dennis Michael Warren. "The Role of Indigenous Tillage Systems in Sustainable Food Production." *Agriculture and Human Values* 8, (1991): 149–155. https://doi.org/10.1007/BF01579667.

Remignon, Herve and Pierre Burgues. "Evolution of Oxidative Stress Markers in Livers of Ducks During Force-Feeding." *Scientific Reports* 13, no. 1046 (2023). https://doi.org/10.1038/s41598-022-27271-y.

Rilke, Rainer Maria. *The Poetry of Rainer Maria Rilke*. United Kingdom: CreateSpace Independent Publishing Platform, 2015.

Romm, Cari. "Rethinking One of Psychology's Most Infamous Experiments." *The Atlantic*. January 28, 2015. https://www.theatlantic.com/health/archive/2015/01/rethinking-one-of-psychologys-most-infamous-experiments/384913/.

Rosenbaum, Emily, and Samantha Friedman. *The Housing Divide—How Generations of Immigrants Fare in New York's Housing Market*. New York: NYU Press, 2006.

Rouge, Melissa. "Semen Collection from Bulls." Colorado State University. Last modified September 2, 2002. http://www.vivo.colostate.edu/hbooks/pathphys/reprod/semeneval/bull.html.

Rozenbaum, Mia. "Do Fish Feel Pain?" Understanding Animal Research. Last modified December 16, 2021. https://www.understandinganimalresearch.org.uk/news/do-fish-feel-pain.

Ruggieri, Amanda. "The Stigma of the Stay-At-Home Dad." British Broadcasting Corporation. Published February 9, 2023. https://www.bbc.com/worklife/article/20230206-are-stay-at-home-dads-becoming-more-common#:~:text=But%20they're%20still%20relatively,fathers%20and%20non%2Dworking%20mothers.

Sager, Gene. "Where's Your Beef From? Grass-fed Beef: Is It Green, Humane, & Healthful?" Natural Life Magazine. Accessed May 2, 2023. https://www.life.ca/naturallife/0812/grass-fed_beef_green_humane_healthful.htm.

Salary.com. "TYSON FOODS INC -CL A Compensation by Company." Accessed July 4, 2023. https://www1.salary.com/TYSON-FOODS-INC-CL-A-Executive-Salaries.html.

Savory, Allan. "How to Fight Desertification and Reverse Climate Change." TED2013. Video, 22:19. https://www.ted.com/talks/allan_savory_how_to_fight_desertification_and_reverse_climate_change/transcript?language=en.

Sayer, Chris. "Troubled Waters | The Women Fighting Sexism in the Fly Fishing Industry." Vice. Published August 1, 2019. https://amuse.vice.com/en_us/article/gyaknm/fly-fishing-sexism.

Scanes, C.G., G. Brant, and M.E. Ensminger. *Poultry Science* 4th Edition (Upper Saddle River, NJ: Pearson Prentice Hall, 2004).

Peter Scarborough, Michael Clark, Linda Cobiac, Keren Papier, Anika Knuppel, John Lynch, Richard Harrington, Tim Key, and Marco Springmann. "Vegans, Vegetarians, Fish-Eaters and Meat-Eaters in the UK Show Discrepant Environmental Impacts." *Nature Food* (2023), https://doi.org/10.1038/s43016-023-00795-w.

Scasta, John Derek, John R. Weir, and Michael C. Stambaugh. "Droughts and Wildfires in Western U.S. Rangelands." *Rangelands* 38, no. 4 (2016): 197–203. https://doi.org/10.1016/j.rala.2016.06.003.

Schorger, A.W. *The Wild Turkey: Its History and Domestication.* Norman: University of Oklahoma Press, 1966. https://d3gxcg0i30gmh1.cloudfront.net/095a-98960757-Wild%20Turkey%20Population%20History%20and%20Overview.pdf?versionId=vOorIBvKXRR1_3HOzqp7bW7Ri8J5gOiA#:~:text=Pre%2DColumbian%20populations%20of%20wild,purposes%20%5B2%2C6%5D.

Science Daily. "Eating Meat Linked to Higher Risk of Diabetes." Published September 5, 2017. https://www.sciencedaily.com/releases/2017/09/170905134506.htm.

Science Daily. "Meat Consumption Contributing to Global Obesity." Published August 16, 2023. https://www.sciencedaily.com/releases/2016/08/160801093003.htm.

Scientific Committee on Animal Health and Animal Welfare. "Welfare Aspects of the Production of Foie Gras in Ducks and Geese." Published December 16, 1998. https://food.ec.europa.eu/system/files/2020-12/sci-com_scah_out17_en.pdf.

SETI Institute. "Earthling Project Launches." Published December 17, 2020. https://www.seti.org/press-release/earthling-project-launches.

Sexton, Chrissy. "Earth, Humans, and All Living Creatures are Made from Stars." Published December 3, 2022. https://www.earth.com/video/earth-humans-and-all-living-creatures-are-made-from-stars/.

Shahbandeh, M. "Total Number of Turkeys Produced in the United States from 1999 to 2021." Statista. Published October 10, 2022. https://www.statista.com/statistics/196092/total-number-of-turkeys-raised-in-the-us-since-1999/.

Simon, Stephanie. "A Killing Floor Chronicle." *Los Angeles Times*, December 8, 2003. https://www.latimes.com/archives/la-xpm-2003-dec-08-na-virgil8-story.html.

Sinclair, Upton. "Chapter III." In *The Jungle*, 1906. Champaign, Illinois: Project Gutenberg, e-book release date 1994. https://www.gutenberg.org/files/140/140-h/140-h.htm.

Singh, R. Paul, and Joe M Regenstein. "Poultry Processing," *Encyclopaedia Britannica*. Last modified August 15, 2018. https://www.britannica.com/technology/poultry-processing.

Skippon, Warren. "The Animal Health and Welfare Consequences of Foie Gras Production." *The Canadian Veterinary Journal* 54, no. 4 (2013): 403. https://www.ncbi.nlm.nih.gov/pmc/articles/PMC3595949/.

Slade, Jessica, and Emma Alleyne. "The Psychological Impact of Slaughterhouse Employment: A Systematic Literature Review." *Trauma, Violence, & Abuse* 24, no. 2 (2021): 429–40. https://doi.org/10.1177/15248380211030243.

Smith, Zadie. *Intimations Six Essays*, 4. London: Penguin Books, 2020.

Soares, Marta C., Rui F. Oliveira, Albert F.H. Ross, Alexandra S. Grutter, and Redouan Bshary. "Tactile Stimulation Lowers Stress in Fish." *Nature Communications* 2, no. 534 (2011). https://doi.org/10.1038/ncomms1547.

Steinfeld, Henning, Pierre Gerber, Tom Wassenaar, Vincent Castel, Mauricio Rosales, and Cees de Haan. *Livestock's Long Shadow, Environmental Issues and Options*. Rome: United Nations Food and Agriculture Organization, 2006. https://www.fao.org/3/a0701e/a0701e.pdf.

Stephen, David. "A Tale of Two Caviars—Caviar and No-Kill Caviar." Marshallberg Farm. Published May 6, 2022. https://thecaviarfarm.com/caviar-and-no-kill-caviar/.

Song, Minyang, Teresa T. Fung, Frank B. Hu, Walter C. Willett, Valter D. Longo, Andrew T. Chan, and Edward L. Giovannucci. "Association of Animal and Plant Protein Intake With All-Cause and Cause-Specific Mortality." *JAMA Internal Medicine* 176, no. 10 (2016): 1453–63. https://doi.org/10.1001/jamainternmed.2016.4182.

Song, Yan, Jorge E. Chavarro, Yin Cao, Weiliang Qiu, Lorelei Mucci, Howard D. Sesso, Meir J. Stampfer, Edward Giovannucci, Michael Pollak, Simin Liu, and Jing Ma. "Whole Milk Intake Is Associated with Prostate Cancer-Specific Mortality Among U.S. Male Physicians." *The Journal of Nutrition* 143, no. 2 (2013): 189–96. https://doi.org/10.3945/jn.112.168484.

Southern Sustainable Agriculture Research and Education. "Artificial Light Sources in Chicken Houses Increase Egg Laying in Hens During Fall and Winter." Published April 5, 2016. https://southern.sare.org/news/artificial-light-sources-in-chicken-houses-increase-egg-laying-in-hens-during-fall-and-winter/.

Souza, Kim. "Tyson Foods Maintains its Top Ranking in Poultry Production." Talk Business & Politics. Published March 20, 2019. https://talk-business.net/2019/03/tyson-foods-maintains-its-top-ranking-in-poultry-production/.

Stone Barns Center. "Back to Grass the Market Potential for U.S. Grass-Fed Beef." Published April 2017. https://www.stonebarnscenter.org/wp-content/uploads/2017/10/Grassfed_Full_v2.pdf.

Storhaug, Christian Løvold, Svein Kjetil Fosse, and Lars T. Fadnes. "Country, Regional, and Global Estimates for Malabsorption in Adults: A Systemic Review and Meta-Analysis." *The Lancet Gastroenterology & Hepatology* 2, no. 10 (2017). https://doi.org/10.1016/S2468-1253(17)30154-1.

Stuesse, Angela, and Nathan T. Dollar. "Who Are America's Meat and Poultry Workers?" Published September 24, 2020. https://www.epi.org/blog/meat-and-poultry-worker-demographics/.

Turner-McGrievy, Gabrielle. Trisha Mandes, and Anthony Crimarco. "A Plant-Based Diet for Overweight and Obesity Prevention and Treatment." *Journal of Geriatric Cardiology* 5, no. 5 (May 2017): 369–74. https://www.ncbi.nlm.nih.gov/pmc/articles/PMC5466943/.

United Egg Producers. "Animal Husbandry Guidelines for U.S. Egg-Laying Flocks." Published 2017. https://uepcertified.com/wp-content/uploads/2021/08/CF-UEP-Guidelines_17-3.pdf.

United Egg Producers. "United Egg Producers Updated Statement on Male Chicks." Published March 25, 2021. https://unitedegg.com/united-egg-producers-updated-statement-on-male-chicks-2/.

United Egg Producers. "U.S. Egg Production and Hen Population." Accessed May 31, 2023. https://unitedegg.com/facts-stats/.

United Nations Environment Programme. "Methane Emissions Are Driving Climate Change. Here's How to Reduce Them." Published August 20, 2021. https://www.unep.org/news-and-stories/story/methane-emissions-are-driving-climate-change-heres-how-reduce-them.

United Nations News. "Conserving Plant Genetic Diversity Crucial for Food Security." Published October 26, 2010. https://news.un.org/en/story/2010/10/357072#:~:text=FAO%20estimates%20that%2075%20per,because%20of%20a%20changing%20climate.

United Poultry Concerns. "Virgil Butler—'Inside Tyson's Hell: Why I Got Out of the Chicken Slaughtering Business.'" YouTube Video, 58:40. Recorded August 2004. https://www.youtube.com/watch?v=2vWqOaP_yX8.

United States Department of Agriculture. "Cage-Free Verification of USDA Graded Shell Eggs." Published December 2023. https://www.ams.usda.gov/sites/default/files/AMSFactSheetCageFreeVerification.pdf.

United States Department of Agriculture. "Does the Label, "Free Range," Pertain to Poultry or Also Other Meats?" Published March 17, 2023. https://ask.usda.gov/s/article/Does-the-label-free-range-pertain-only-to-poultry-or-also-to-meats.

United States Department of Agriculture. "USDA Coexistence Fact Sheets Corn." Published February 2015. https://www.usda.gov/sites/default/files/documents/coexistence-corn-factsheet.pdf.

United States Department of Agriculture. "USDA Coexistence Fact Sheets Soybeans." Published February 2015. https://www.usda.gov/sites/default/files/documents/coexistence-soybeans-factsheet.pdf.

United States Department of Agriculture Agricultural Marketing Service. "Questions and Answer—USDA Shell Grading Service." Published October 1, 2015. https://www.ams.usda.gov/publications/qa-shell-eggs.

United States Department of Agriculture Animal and Plant Health Inspection Service. "Program Data Report G—2021 Animals Dispersed/Killed or Euthanized/ # Burrows/Dens Removed or Destroyed/ Free or Relocated." Accessed June 19, 2023. https://www.aphis.usda.gov/aphis/ourfocus/wildlifedamage/SA_Reports/SA_PDRs.

United States Department of Agriculture Animal and Plant Health Inspection Service. "Protecting Livestock from Predators." Last modified November 1, 2022. https://www.aphis.usda.gov/aphis/ourfocus/wildlifedamage/operational-activities/SA_Livestock/CT_Protecting_Livestock_Predators.

United States Department of Agriculture Economic Research Service. "Cattle & Beef: Sector at a Glance." Last modified September 26, 2022. https://www.ers.usda.gov/topics/animal-products/cattle-beef/sector-at-a-glance/.

United States Department of Agriculture Economic Research Service. "Farm Structure and Contracting." Last modified March 8, 2022. https://www.ers.usda.gov/topics/farm-economy/farm-structure-and-organization/farm-structure-and-contracting/.

United States Department of Agriculture Economic Research Service. "The Meatpacking Industry in Rural America During the COVID-19 Pandemic." Last modified May 13, 2021. https://www.ers.usda.gov/covid-19/rural-america/meatpacking-industry/.

United States Department of Agriculture Economic Research Service. "Poultry Sector at a Glance." Accessed July 4, 2023. https://www.ers.usda.gov/topics/animal-products/poultry-eggs/sector-at-a-glance/.

United States Department of Agriculture Economics, Statistics and Market Information System. "Livestock Slaughter." Accessed May 26, 2023. https://usda.library.cornell.edu/concern/publications/rx913p88g?locale=en.

United States Department of Agriculture Economics, Statistics and Market Information System. "Poultry Slaughter." Accessed May 26, 2023. https://usda.library.cornell.edu/concern/publications/3197xm04j?locale=en.

United States Department of Agriculture Farm Service Agency. "Crop Acreage Data." Last modified August 2022. https://www.fsa.usda.gov/news-room/efoia/electronic-reading-room/frequently-requested-information/crop-acreage-data/index.

United States Department of Agriculture Food Safety and Inspection Service. "Constituent Update—February 23, 2018, FSIS Criteria for Consideration of Waiver Requests from Young Chicken Slaughter Establishment to Operate at Line Speeds Up to 175 Birds Per Minute." Accessed July 4, 2023. https://www.fsis.usda.gov/news-events/news-press-releases/constituent-update-february-23-2018-0.

United States Department of Agriculture National Agricultural Library. "Animal Welfare Act." Accessed July 3, 2023. https://www.nal.usda.gov/animal-health-and-welfare/animal-welfare-act#:~:text=The%20Animal%20Welfare%20Act%20(AWA,USDA%20%2C%20APHIS%20%2C%20Animal%20Care.

United States Department of Agriculture National Agricultural Library. "Humane Methods of Slaughter Act." Accessed July 3, 2023. https://www.nal.usda.gov/animal-health-and-welfare/humane-methods-slaughter-act.

United States Department of Agriculture National Agricultural Statistics Service. "Poultry Slaughter, Total Live Weight, and Average Live Weight by Type." Published January 25, 2023. https://downloads.usda.library.cornell.edu/usda-esmis/files/3197xm04j/4j03f9428/ht24xw327/psla0123.pdf.

United States Department of Agriculture National Institute of Food and Agriculture. "Family Farms." Accessed June 27, 2023. https://www.nifa.usda.gov/grants/programs/family-small-farm-program/family-farms#:~:text=USDA%20classifies%20family%20farms%20as,as%20farms%20with%20hired%20managers%E2%80%9D.

United States Department of Agriculture Research, Education & Economics Information System. "Maximization of Laying Hen Performance, Welfare, Economic Return, and Egg Quality and Safety." Published September 30, 2013. https://reeis.usda.gov/web/crisprojectpages/0152013-maximization-of-laying-hen-performance-welfare-economic-return-and-egg-quality-and-safety.html.

United States Department of the Interior Bureau of Land Management. "Livestock Grazing on Public Lands." Accessed July 4, 2023. https://www.blm.gov/programs/natural-resources/rangelands-and-grazing/livestock-grazing.

United States Environmental Protection Agency Global Methane Initiative. "Importance of Methane." Last modified May 22, 2023. https://www.epa.gov/gmi/importance-methane#:~:text=Methane%20is%20more%20than%2025,due%20to%20human%2Drelated%20activities.

United States Food and Drug Administration Center for Veterinary Medicine. "2018 Summary Report on Antimicrobials Sold or Distributed for Use in Food-Producing Animals." Published December, 2019. https://www.fda.gov/media/133411/download.

United States Government Accountability Office. "Workplace Safety and Health: Additional Data Needed to Address Continued Hazards in the Meat and Poultry Industry." Published April 25, 2016. https://www.gao.gov/products/gao-16-337.

United States Government Printing Office. "Animal Enterprise Terrorism Act." Accessed May 26, 2023, https://www.govinfo.gov/content/pkg/PLAW-109publ374/html/PLAW-109publ374.htm.

UN Women—United Nations Entity for Gender Equality and the Empowerment of Women. "Transforming Economies, Realizing Rights." Published 2015. https://progress.unwomen.org/en/2015/pdf/UNW_progressreport.pdf.

Victor, Karen, and Antoni Barnard. "Slaughtering for a Living: A Hermeneutic Phenomenological Perspective on the Well-Being of Slaughterhouse Employees." *International Journal of Qualitative Studies on Health and Well-being* 11, no. 1 (2016): 30266. https://doi.org/10.3402/qhw.v11.30266.

Virginia Business. "Equilar Virginia CEO Pay Report." Accessed July 4, 2023. https://www.virginiabusiness.com/pdf/1012ceopaychart.pdf.

Wang, Y., and M. A. Beydoun. "Meat Consumption Is Associated with Obesity and Central Obesity Among US Adults." *International Journal of Obesity* 33 (2009): 621–28. https://doi.org/10.1038/ijo.2009.45.

Warriss, P.D., E.A. Bevis, and S.N. Brown. "Time Spent by Broiler Chickens in Transit to Processing Plants." *The Veterinary Record* 127, no. 25–26 (1990): 617–19. https://pubmed.ncbi.nlm.nih.gov/2089775/.

The Washington Post. "Chickens, Turkeys Being Boiled Alive as Processing Lines Speed Up." Last modified August 28, 2017. https://www.dailynews.com/2013/10/30/chickens-turkeys-being-boiled-alive-as-processing-lines-speed-up/.

Wasley, Andrew, Christopher D. Cook, and Natalie Jones. "Two Amputations a Week: The Cost of Working in a US Meat Plant, *The Guardian*, July 5, 2018. https://www.theguardian.com/environment/2018/jul/05/amputations-serious-injuries-us-meat-industry-plant#:~:text=US%20meat%20workers%20are%20already,to%20suffer%20repetitive%20strain%20injuries.

Water Footprint Network. "Do You Know How Much Water Was Used to Grow Your Food and to Produce Your Clothes and the Things You Buy?" Accessed July 4, 2023. https://www.waterfootprint.org/time-for-action/what-can-consumers-do/.

Weber, Christopher L., and H. Scott Matthews. "Food-Miles and the Relative Climate Impacts of Food Choices in the United States." *Environmental Science & Technology* 42, no. 10 (2008): 3508–13. https://doi.org/10.1021/es702969f.

Weingus, Leigh. "Are Pasture-Raised Eggs Worth the Splurge? Nutritionists Weigh In." Huffington Post. Published August 2, 2021. https://www.huffpost.com/entry/are-pasture-raised-eggs-healthier_l_61005ad3e4b000b997dbbbcc.

Weisse, Mikaela, and Elizabeth Goldman. "Just Seven Commodities Replaced an Area of Forest Twice the Size of Germany between 2001 and 2015." World Resources Institute. Published February 11, 2021. https://www.wri.org/insights/just-7-commodities-replaced-area-forest-twice-size-germany-between-2001-and-2015.

Whitehead, C.C. "Skeletal Disorders in Laying Hens: The Problem of Osteoporosis and Bone Fractures." In *Welfare of the Laying Hen*, edited by G.C. Perry, 259–78. Wallingford, Oxfordshire: CABI Publishing, 2004.

Whitlock, B.K., E.A. Coffman, J.F. Coetzee, J.A. Daniel. "Electroejaculation Increased Vocalization and Plasma Concentrations of Cortisol and Progesterone, but Not Substance P, in Beef Bulls." *Theriogenology* 78 (2012): 737–46. https://doi.org/10.1016/j.theriogenology.2012.03.020.

Woodson, Mariah. "The Best We Can Be: How Toxic Masculinity Creates a Second Inescapable Situation for Inmates." *Public Interest Law Reporter* 24, no. 2 (2019): 2. https://lawecommons.luc.edu/cgi/viewcontent.cgi?article=1569&context=pilr.

Xu, Rongbin, Pei Yu, Michael J. Abramson, Fay H. Johnston, Jonathan M. Samet, Michelle L. Bell, Andy Haines, Kristie L. Ebi, Shanshan Li, and Yuming Guo. "Wildfires, Global Climate Change, and Human Health." *The New England Journal of Medicine* 383 (November 26, 2020): 2173–81. https://doi.org/10.1056/NEJMsr2028985.

Xu, Xiaming, Prateek Sharma, Shijie Shu, Tzu-Shun Lin, Philippe Ciais, Francesco N. Tubiello, Pete Smith, Nelson Campbell, and Atul K. Jain. "Global Greenhouse Gas Emissions from Animal-Based Food Are Twice Those of Plant-Based Foods." *Nature Food* 2 (September 2021): 724–32. https://www.nature.com/articles/s43016-021-00358-x.

Yong, Ed. "Not a Hug Hormone—Fish Version of Oxytocin Acts as Social Spotlight." National Geographic. Published October 10, 2012. https://www.nationalgeographic.com/science/article/fish-oxytocin-isotocin-social-spotlight.

You, Wenpeng, and Maciej Henneberg. "Meat Consumption Providing a Surplus Energy in Modern Diet Contributes to Obesity Prevalence: An Ecological Analysis." *BMC Nutrition* 2, no. 22 (2016). https://doi.org/10.1186/s40795-016-0063-9.

Zhuang, Pan, Fei Wu, Lei Mao, Fanghuan Zhu, Yiju Zhang, Xiaoqian Chen, Jingjing Jiao, and Yu Zhang. "Egg and Cholesterol Consumption and Mortality from Cardiovascular and Different Causes in the United States: A Population-Based Cohort Study." *PLoS Medicine* 18, no. 2 (February 2021). https://doi.org/10.1371/journal.pmed.1003508.

Zielinski, Sarah. "Climate Change Will Accelerate Earth's Sixth Mass Extinction." *Smithsonian Magazine*, April 30, 2015. https://www.smithsonianmag.com/science-nature/climate-change-will-accelerate-earths-sixth-mass-extinction-180955138/.

Zimmerman Jones, Andrew. "What Is the Visible Light Spectrum?" ThoughtCo. Last modified February 14, 2020. https://www.thoughtco.com/the-visible-light-spectrum-2699036.

Zippia. "Homemaker Demographics and Statistics in the US." Published 2021. https://www.zippia.com/homemaker-jobs/demographics/.

Zippia. "Homemaker Salary." Last modified April 6, 2023. https://www.zippia.com/homemaker-jobs/salary/.

ZipRecruiter. "Poultry Farm Worker Salary." Accessed June 5, 2023. https://www.ziprecruiter.com/Salaries/Poultry-Farm-Worker-Salary.

zur Hausen, Harald. "Red Meat Consumption and Cancer: Reasons to Suspect Involvement of Bovine Infectious Factors in Colorectal Cancer." *International Journal of Cancer* 130, no. 11 (June 2012): 2475–83. https://doi.org/10.1002/ijc.27413.

About the Contributors

Rosemary Alles ("Supermoon") graduated in computer science and mathematics from the University of Toronto. She has a master's degree in Geographic Information Systems from Penn State; her work as a software engineer for astronomical observatories spanned several years, during which she worked for the Keck and CFHT telescopes in Hawaii, Caltech in Southern California, and NASA/Ames (USRA) in Silicon Valley. Today, her work is at the intersection of disenfranchised animal and human communities. She believes that creative disruptions of outdated modalities of conservation will lead to sustainable solutions in The Global March for Elephants and Rhino's efforts to liberate both animals and humans.

Chloe Hanson ("Dairy Pride") holds a PhD from the University of Tennessee, where she studied under Poet Laureate Joy Harjo. Her poetry has been featured in several journals online and in print, including *The Rumpus*, *Cimarron Review*, *Third Coast*, *Glass: A Journal of Poetry*, and *Gargoyle Literary Magazine*. Her collection of poetry, *Bluebeard's Key*, uses fairy tales to bear witness to the parallel and intersecting oppressions of women and nonhuman animals living in Western patriarchal societies.

pattrice jones ("The Persistence of Pigeons") is a cofounder of VINE Sanctuary, an LGBTQ-led farmed animal refuge caring for more than 600 nonhuman animals while working for social and environmental justice as well as animal liberation. Two aviaries at the sanctuary shelter scores of pigeons rescued from meat markets, pigeon shoots, pigeon racing, and animal experimentation, and the sanctuary grounds serve as a home base for flocks of rewilded descendants of such survivors.

Mona Seymour, PhD ("A Landscape Rewritten") is an associate professor of urban and environmental studies at Loyola Marymount University in Los Angeles, CA. She is broadly interested in the place of veganism and animal liberation politics in the agrifood movement. Her scholarship focuses on vegan or animal-free organic agriculture and has recently included research on the human dimensions of veganic farming in the United States, in collaboration with Alisha Utter of the University of Vermont. She is also involved in organizing around US veganic farming, including maintaining a map of North American veganic farms in order to facilitate connections within this agricultural community.

About the Author

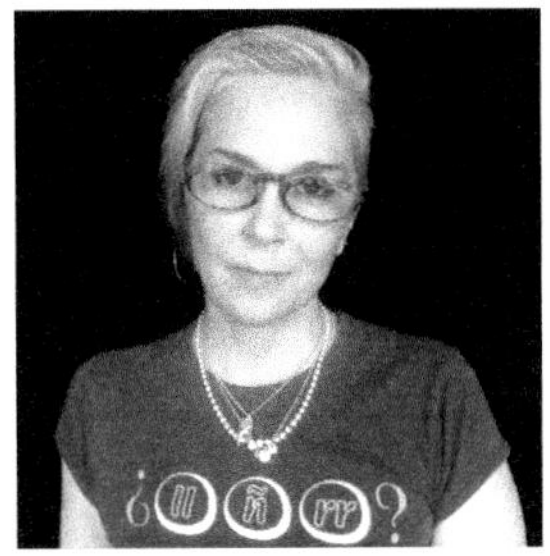

Isabella La Rocca González is an artist and activist based in the United States. Her work is part of a long tradition in art and photography: to bring to light and find beauty in the disregarded, hidden, or unconscious. As a first-generation American, she strives to reconcile values from her Indigenous Mexican roots with her European heritage. Her photographs have been exhibited internationally, including a solo show at the Center for Photography in Woodstock, NY. Her creative nonfiction works have been published in various venues, including the *Women Eco Artist Dialog Magazine* and the *Everything from Nothing Journal*. Her screenplay, *Fugue 9*, was chosen as a finalist for the 2008 Sundance Screenwriter's Lab. She received her BA in fine arts from the University of Pennsylvania, and her MFA in Photography from Indiana University. After thirty years teaching art and photography on the post-secondary level, she has left academia to devote herself full time to her art practice and to growing a native pollinator garden.

About the Publisher

Lantern Publishing & Media was founded in 2020 to follow and expand on the legacy of Lantern Books—a publishing company started in 1999 on the principles of living with a greater depth and commitment to the preservation of the natural world. Like its predecessor, Lantern Publishing & Media produces books on animal advocacy, veganism, religion, social justice, humane education, psychology, family therapy, and recovery. Lantern is dedicated to printing in the United States on recycled paper and saving resources in our day-to-day operations. Our titles are also available as ebooks and audiobooks.

To catch up on Lantern's publishing program, visit us at www.lanternpm.org.

facebook.com/lanternpm
instagram.com/lanternpm
tiktok.com/@lanternpmofficial